The Muscular & Skeletal Systems

Titles in the series

The Muscular & Skeletal Systems

The Circulatory & Respiratory Systems

The Digestive & Urinary Systems

The Reproductive & Nervous Systems

THE HUMAN BODY & THE ENVIRONMENT

How our surroundings affect our health

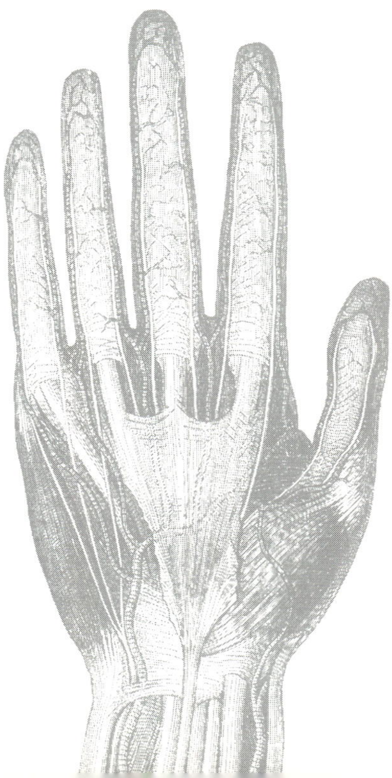

The Muscular & Skeletal Systems

GREENWOOD PRESS
Westport, Connecticut · London

Library of Congress Cataloging-in-Publication Data

Creative Media Applications
The human body & the environment/how our surroundings affect our health
p. cm.—(Middle school reference)
Includes bibliographical references and index.
Contents: v. 1. The muscular & skeletal systems—v. 2. The circulatory & respiratory systems—v. 3. The digestive & urinary systems—v. 4. The reproductive & nervous systems.
ISBN 0-313-32558-8 (set: alk. paper)—0-313-32559-6 (v.1)—0-313-32560-X (v.2) —0-313-32561-8 (v.3)—0-313-32562-6 (v.4)
1. Human physiology—Juvenile literature. [1. Human physiology. 2. Environmental health.] I. Title: Human body and environment. II. Series.
QP37.H8925 2003
612—dc21 2002035217

British Library Cataloguing in Publication Data is available.

Library of Congress Catalog Card Number: 2002035217

ISBN: 0–313–32558–8 (set)
0–313–32559–6 (vol. 1)
0–313–32560–X (vol. 2)
0–313–32561–8 (vol. 3)
0–313–32562–6 (vol. 4)

First published in 2003

Greenwood Press, 88 Post Road West, Westport, CT 06881
An imprint of Greenwood Publishing Group, Inc.
www.greenwood.com

Printed in the United States of America

The paper used in this book complies with the Permanent Paper Standard issued by the National Information Standards Organization (Z39.48–1984).

10 9 8 7 6 5 4 3 2 1

A Creative Media Applications, Inc. Production
WRITER: Robin Doak
DESIGN AND PRODUCTION: Fabia Wargin Design, Inc.
EDITOR: Matt Levine
COPYEDITOR: Laurie Lieb
PROOFREADER: Barbara Francis
ASSOCIATED PRESS PHOTO RESEARCHER: Yvette Reyes
CONSULTANT: Michael Windelspecht

PHOTO CREDITS:
Cover: (*top*) ©Photodisc and (*bottom*) ©Veer
AP/Wide World Photographs *pages* vi, 4, 16, 21, 27, 32, 34, 38, 41, 44, 53, 57, 63, 66, 68, 71, 72, 77, 79, 81, 95, 103, 104, 113, 115, 117, 121, 122, 125
© Bob Krist/CORBIS *page* ix
© Buck/Custom Medical Stock Photo *page* 11
© Potokar/Custom Medical Stock Photo *page* 14
© B.S.I.P./Custom Medical Stock Photo *page* 37
© Birmingham/Custom Medical Stock Photo *page* 50
© Custom Medical Stock Photo *pages* 61, 98
© Logical Images/Custom Medical Stock Photo *pages* 74, 99, 106
© NMSB/Custom Medical Stock Photo *pages* 84, 92, 97
© Croce/Custom Medical Stock Photo *page* 85
© Mike Buxton; Papilio/CORBIS *page* 108

ILLUSTRATION CREDITS:
© Lifeart *pages* 1, 3, 6, 7, 9, 10, 13, 19, 31, 39, 45, 47, 88, 119

CONTENTS

INTRODUCTION . . . vii

1 THE MUSCULOSKELETAL SYSTEM . . . 1

2 BONE DISORDERS . . . 13

3 MUSCLE DISORDERS . . . 31

4 JOINT AND CONNECTIVE TISSUE DISORDERS . . . 39

5 HOW THE ENVIRONMENT AFFECTS THE MUSCULOSKELETAL SYSTEM . . . 59

6 THE SKIN . . . 87

7 SKIN DISORDERS . . . 91

8 HOW THE ENVIRONMENT AFFECTS THE SKIN . . . 101

9 KEEPING THE MUSCULOSKELETAL SYSTEM AND SKIN HEALTHY . . . 119

GLOSSARY . . . 127

BIBLIOGRAPHY . . . 129

INDEX . . . 130

INTRODUCTION

LIVING IN A RISKY ENVIRONMENT

Environmental health is usually defined as the human body's reaction to conditions that we have little or no control over. Environmental health concerns how our bodies react to both natural and human-made substances. Sunlight and bacteria are two of the natural agents that can affect our health. Chemicals, air pollution, and water pollution are human-made agents that can also affect the way that our bodies work and function.

Environmental health can also be affected by medical problems that result from personal lifestyle choices, as well as conditions in the workplace. That's because certain decisions can affect our health. For example, the decision to exercise and eat a healthy diet will improve a person's health. Smoking cigarettes and drinking alcohol, on the other hand, could cause serious medical problems. Repetitive stress injuries can be the result of job-related activities or sports injuries.

Children are especially at risk for environmental diseases. According to the World Health Organization (WHO), 3 million children under the age of five die each year from diseases caused by unclean drinking water, indoor air pollution, and other environmental agents. Children are not "little adults." Their bodies are still growing and developing. These growing bodies are not yet equipped to absorb and remove chemicals and other pollutants in the same way that adult bodies can. Pollution and other environmental agents may have a longer-lasting impact on children than on full-grown adults.

Opposite:
Air pollution in the form of smog darkens the atmosphere along one of the many freeways in Los Angeles.

THE ENVIRONMENT AND THE BODY

Every day, each and every one of us is exposed to environmental agents in the world around us. These environmental agents include chemical, biological, and physical substances. Chemical agents include any of the thousands of natural and human-made chemicals in the world. They also include pesticides, food additives, and pollutants in the air, soil, and water. Biological factors include any of the many disease-causing bacteria, viruses, and other microorganisms that we unwittingly come in contact with. These organisms might be in our food, our water, or the air. Physical agents include workplace conditions, traffic accidents, noise pollution, radiation, and excessive heat or cold in the environment.

Environmental agents can affect the health of the entire body, including the musculoskeletal system. Radiation, chemical exposure, and biological agents can all increase the risk of serious muscle, bone, and joint disorders. These disorders include osteoporosis, spina bifida, tetanus, and Lyme disease. Nutritional deficiencies can also cause musculoskeletal problems.

The skin can also be negatively affected by environmental agents. Skin cancer in the United States, most often caused by the sun's ultraviolet radiation, has increased by about 7 percent each year since 1997. According to the Centers for Disease Control and Prevention (CDC), more than 1 million Americans were diagnosed with some type of skin cancer in 2002. Other environmental agents that can irritate and damage the body's protective layer include bacteria, viruses, chemicals, smoking, and pollution.

Although you cannot avoid all contact with environmental agents that can negatively affect your health, it is important to know how to keep the body healthy. Diet, exercise, and understanding how environmental conditions affect you are all important steps that you can take to keep your body as healthy as possible.

HOW TO USE THIS BOOK

Each volume of this series covers a number of body systems and the diseases—environmental and otherwise—that can affect them. Each volume is organized to explain the basic anatomy of each system, including its structure and function. This volume includes the musculoskeletal system and the skin.

Environmental Health Factors

The National Institute of Environmental Health Sciences (NIEHS) lists the following environmental factors that can affect a person's health:

- foods and other nutrients
- physical agents such as heat and *radiation,* which is energy in the form of rays or particles
- social and economic factors that affect health and behavior
- lifestyle choices, including substance abuse and unprotected sexual intercourse
- artificial and naturally occurring chemicals

An asthma sufferer holds her many medications as she stands within view of a paper mill. Asthma is one of many adverse health conditions that are affected by nutrition, physical agents, social and economic factors, lifestyle, and artificially and naturally occurring chemicals in the environment.

The nonenvironmental and the environmental medical conditions that affect each body system are covered in separate chapters. Each disease listing includes the condition's causes, symptoms, and treatments or cures. Additional information may include diagnostic tools, statistics, and historical information.

The final chapter in each volume offers tips on how to keep the body healthy. A glossary offers definitions of some words in the book. Further, pronunciations of some of the more difficult words in the book appear throughout. For a complete list of pronunciations, however, consult a medical dictionary. The Index can easily be used to find a particular disorder.

Finally, the volumes are not meant to be used for self-diagnosis of medical conditions. Those who have health problems should always consult a medical professional.

Note: All metric conversions in this book are approximate.

1

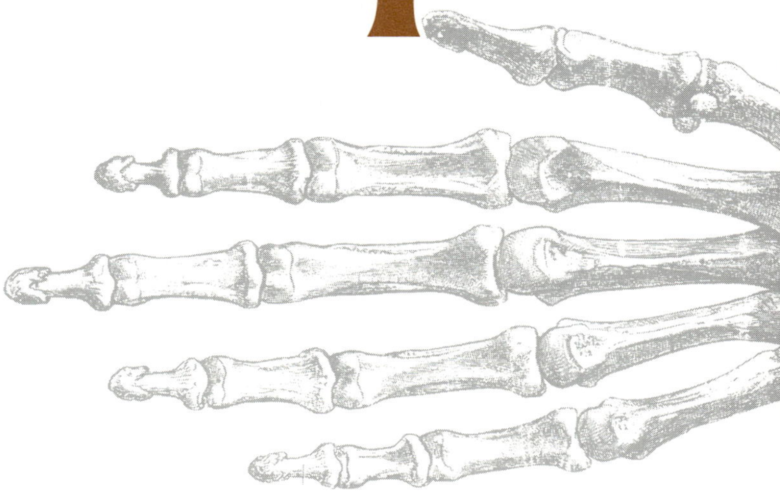

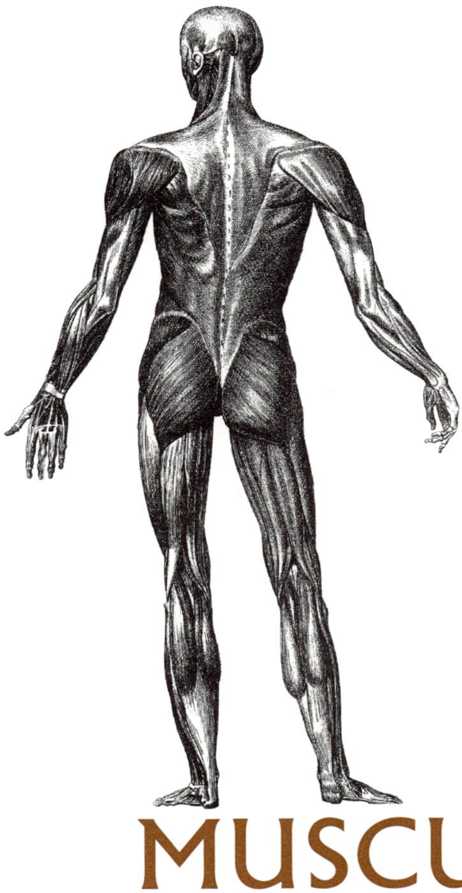

THE MUSCULOSKELETAL SYSTEM

The *musculoskeletal* (mus-kyoo-loh-SKEL-uh-tul) *system* is made up of the bones, muscles, joints, and connective tissues in the body. This system is the body's framework. The musculoskeletal system provides strength and support, allowing the body to move.

THE SKELETON

The human body is made up of about 206 bones. Together, these bones are called the skeleton. The bones provide the structure and form of the body. Without bones, our bodies would be big blobs without shape.

Bones are one of the strongest natural substances on Earth. Bones are strong enough to support the weight of the entire body. Yet they are light enough to permit humans to stand upright, walk around, and perform a wide variety of actions.

What other important roles do the bones play?

- Bones protect the internal organs from harm. The skull, for example, protects the brain, while the ribs protect the heart, lungs, spleen, stomach, and liver.
- Bones provide a place for muscles to attach.
- Bones are a storehouse for minerals and fats. In fact, more than 90 percent of the body's calcium and phosphorous is stored inside bones. Bones release these important nutrients when the body needs them.
- Bone marrow inside some bones produces red and white blood cells.

WHAT'S IN A BONE?

Most people think of bones as the dead, dried-up matter that a skeleton is made of. However, the bones inside the human body consist of a combination of living and nonliving materials. These materials include minerals, protein, water, and living cells.

Two of the most important minerals in bones are calcium and phosphorous. They help make bones hard. The protein in bones, called *collagen* (KAHL-uh-jen), makes them flexible. Blood vessels and nerves run through bones, supplying the bones with the nourishment that they need to stay healthy.

A bone is made up of several different parts.

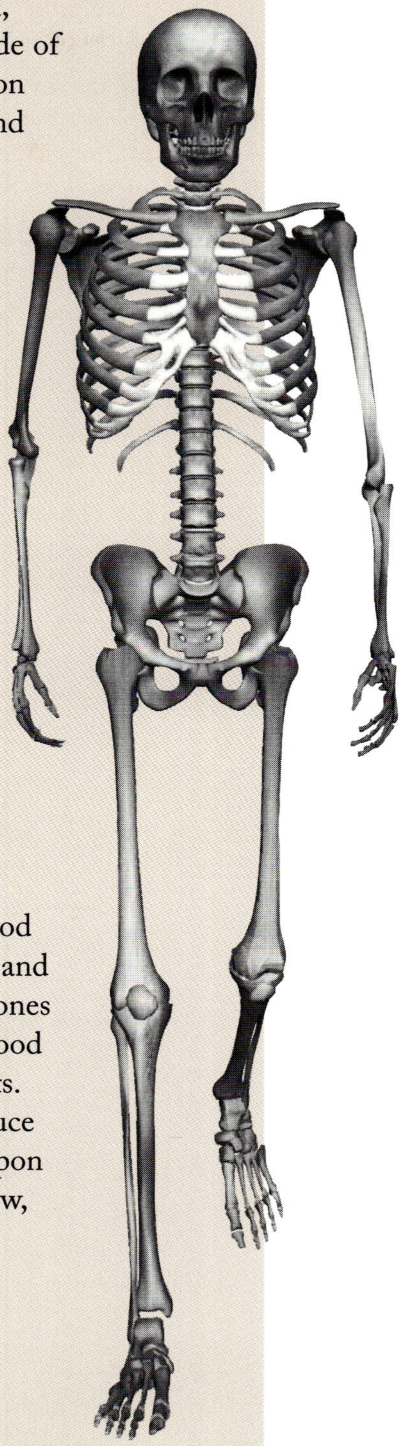

- The *periosteum* (peh-ree-AH-stee-um) is a thin, tough, protective membrane on the outside of the bone surface. It is found everywhere on bones except at the ends. Blood vessels and nerves run through the periosteum.
- *Compact bone,* also called cortical bone, is found beneath the periosteum. This hard, dense outer layer gives bones their super strength. Tiny holes in the bone are the pathway for more nerves and blood vessels.
- Also called cancellous bone, *spongy bone* is found inside compact bone. Spongy bone is much lighter than compact bone. It is filled with many small spaces inside, just like a sponge. Like compact bone, this type of bone also contains nerves and blood vessels.
- The *marrow cavity* is the space in the center of such long bones as the *femur* (thighbone) and the *humerus* (upper arm bone). Bone marrow is found inside the marrow cavity.
- *Marrow* is a type of tissue that makes blood cells. There are two types of marrow: red and yellow. Red bone marrow, found in flat bones and the ends of long bones, makes red blood cells, white blood cells, and blood platelets. Red bone marrow has the ability to produce millions of blood cells daily, depending upon the needs of the body. Yellow bone marrow, found in long bones, stores fat cells.

Talking Bones

Did you know that bones can tell tales? Scientists called *physical anthropologists* (an-throh-PAHL-oh-jists) study the remains of ancient people to learn how humans developed through the ages. By looking at the bones of people who have been dead for thousands of years, these scientists can learn about the diet, diseases, and deaths of these people from another time.

Forensic anthropologists are scientists who examine modern bones in order to solve murders and other crimes. In some criminal cases, bones are the only clues that are left behind to study. Forensic anthropologists use the bones to determine a victim's age, sex, and ancestry. This information could be the key to unraveling a mystery or solving a crime.

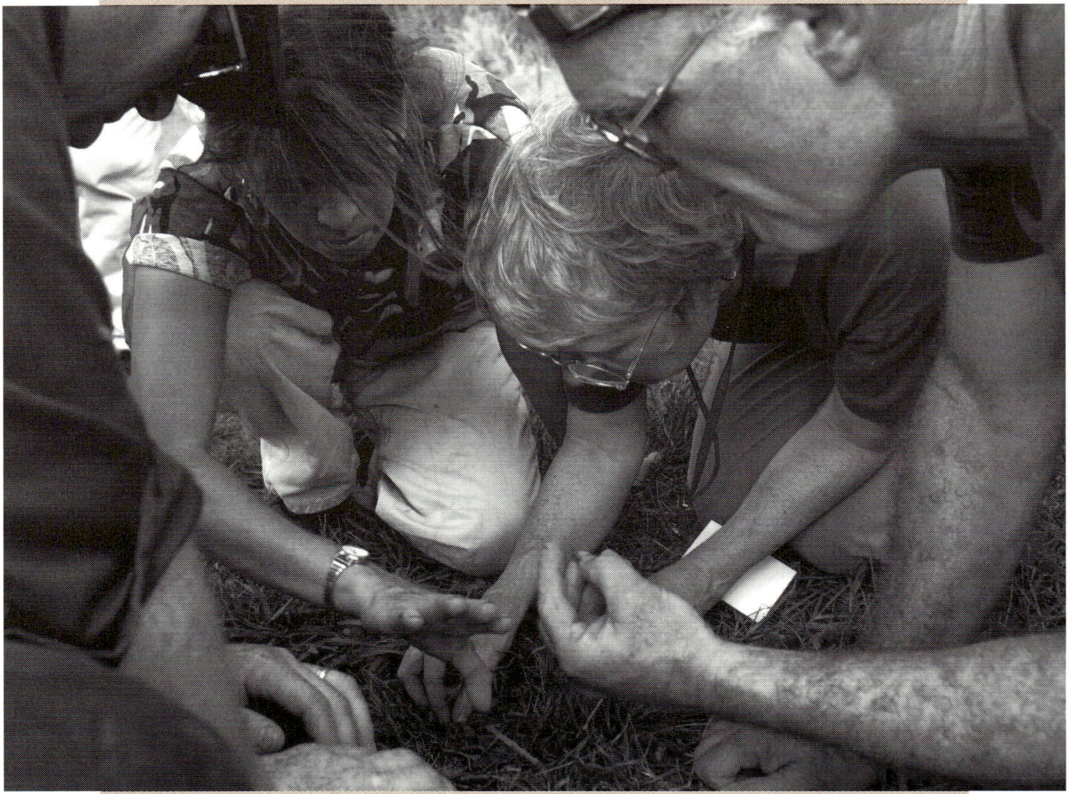

Forensic scientists examine human bones near the Los Pumas military base in Panama City, Panama. They are searching for evidence of murders committed by the military dictatorships that ruled Panama from 1968 to 1989.

Bones are constantly changing and growing. When humans are born, their bones are made up mostly of cartilage. As they get older, the cartilage is replaced with hard, strong, bone cells. This process is called *ossification.* Most humans have a fully developed skeletal system by the age of twenty-five.

Because bones are growing during childhood, it is especially important for kids to do the right things to make sure their bones develop properly. To have healthy bones, humans need to eat a healthy diet rich in vitamins and minerals. Calcium is especially important. All types of physical activity also help bones be as strong as they can be.

Oh, Baby!

Did you know that infants are born with about 350 bones? That's a lot more than the average 206 bones that a grown-up has. So where do these "extra" bones go? As babies grow, many of their bones fuse together to form single bones.

Even in adults, bones continue to break down and rebuild themselves. Bone-building cells, located in the end portions of long bones, are called osteoblasts (AH-stee-oh-blasts). Cells that break down bone are called osteoclasts (AH-stee-oh-klasts). Bones even work to repair themselves after a fracture or other injury.

TYPES OF BONES

The human skeleton is divided into two sections: the *axial* (AK-see-ul) *skeleton* and the *appendicular* (aa-pen-DIK-yuh-ler) *skeleton.* The axial skeleton runs lengthwise down the center of the body. The axial skeleton consists of the skull, ribs, and spine. The appendicular skeleton is made up of the bones that attach to the axial skeleton. It includes the arms, legs, shoulder bones, and pelvis.

Bones come in many different shapes and sizes. Some bones are long and large: The femur, or thighbone, makes up about one-fourth of a person's height. The femur is also the largest, heaviest, and strongest bone in the body. Other bones are tiny: The *stapes* (STAY-peez), or stirrup, a bone in the ear, is only about 0.1 inches (0.25 centimeters) long.

The bones of the body can be sorted into one of four groups: *Long bones* are rod-shaped bones that support weight and bear stress. Legs, arms, and fingers are all long bones. Cube-shaped *short bones* support weight and allow fine motor movements. These types of bones are found in the ankles and the wrists. *Flat bones* are made up of a layer of spongy bone enclosed by two layers of compact bone. These types of bones offer protection and support. Ribs, shoulder blades, and the breastbone are all flat bones. Bones that are not long, short, or flat are *irregular bones.* They include the vertebrae, which are the bones in the spine, as well as some bones in the ear.

Fascinating Skeleton Facts

- An adult's skeleton weighs about 20 pounds (9 kilograms).
- The word *skeleton* comes from the Greek word meaning "dried up."
- One of the last bones to harden in the human body is the breastbone.
- About 70 percent of bone is made up of inorganic minerals. The other 30 percent is made up of proteins and other organic matter.
- The *hyoid* (HYE-oyd) *bone,* located below the lower jaw, is the only bone in the human body that is not connected to another bone. This U-shaped bone supports the tongue and attaches to muscles that allow humans to swallow and speak.

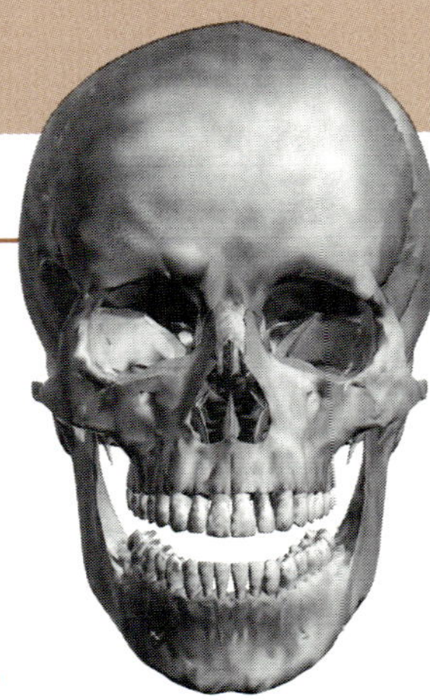

Teeth

Teeth, although hard and bonelike, are not really bones at all. Instead, your teeth are *dentin,* a hard substance that is made up mostly of calcium and collagen. Each human has two sets of teeth in a lifetime. The first set, called *deciduous* (dih-SID-yoo-uss) *teeth,* begins growing in when a child is about seven months old. These twenty small teeth are also known as "baby teeth" or "milk teeth." They start to fall out around the age of six, when the second set of teeth, called *permanent teeth,* begins to push its way through the gums.

By the time children are twelve, they should have nearly all of their adult teeth. The final teeth to appear are the wisdom teeth. Most people get their wisdom teeth by the age of twenty-one. These teeth are not needed for chewing and sometimes crowd the mouth. As a result, wisdom teeth are often removed. The average adult has thirty-two teeth: sixteen in the upper jaw and sixteen in the lower jaw.

MUSCLES

Muscles really get things moving! Muscles are made up of small, thick bundles of fibers. The fibers are formed out of muscle cells. Working with bones, muscles allow people to lift, run, speak, and sit. Even when the human body is resting, the muscles still have a job to do. They remain tense, giving form to the body and keeping bones in the right positions.

Muscles also power movements that we can't see. They move food from our throats to our stomachs. They help blood circulate through the body, and they keep the heart beating. Like bones, muscles need good nutrition and exercise to keep them going strong.

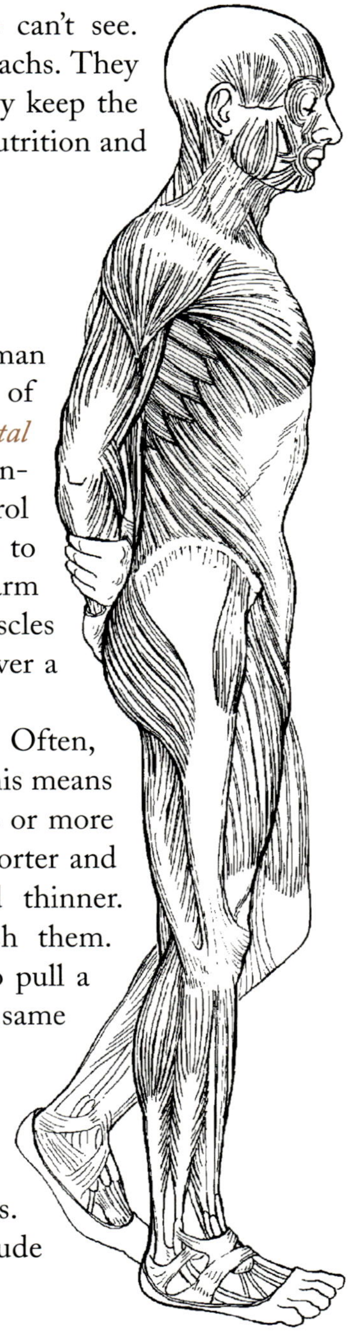

TYPES OF MUSCLES

There are three types of muscles in the human body: skeletal, smooth, and cardiac. About 650 of the 700 muscles in the human body are *skeletal muscles.* Skeletal muscles are also known as voluntary muscles because we can voluntarily control their movement. For example, humans choose to use skeletal muscles when they flex their arm muscles or run laps around a track. Skeletal muscles are usually attached from one end of a bone, over a joint, to the end of another bone.

Skeletal muscles work in pairs or groups. Often, these muscle pairs or groups are *antagonistic.* This means that they work against one another. While one or more muscles in the pair or group *contract,* or get shorter and thicker, the others *relax,* or get longer and thinner. Muscles can pull bones, but they can't push them. Without other muscles in the pair or group to pull a bone back, the bone would remain in the same position forever.

Smooth muscles power movements that we don't control. For this reason, they are also called involuntary muscles. The body's nervous system automatically controls the smooth muscles. Functions powered by these muscles include breathing, digestion, and circulation.

Cardiac (KAR-dee-ak) *muscle* is a special type of muscle found only in the heart. Like smooth muscles, cardiac muscle works automatically. The cardiac muscle of the heart, also called the *myocardium* (mye-oh-KAR-dee-um), constantly contracts and then relaxes, forcing blood through the heart. These contractions create the body's distinctive heartbeat. Most human hearts beat about seventy times in one minute.

Fascinating Muscle Facts

- Muscles are 75 percent water.
- The largest muscle in the human body is the *gluteus maximus.* (GLOO-tee-uss MAKS-ih-muss)—your rear end! Believe it or not, this muscle is also one of the strongest.
- The smallest muscles in the human body are in the ear. They help the tiny bones in your ears move.
- The face has about thirty different muscles. These muscles move when you smile, frown, laugh, or cry.
- The hand has about twenty different muscles, which allow for a wide variety of hand movements.
- Muscles make up about half of the body's entire mass.

JOINTS

A joint is the place where two bones meet. Joints make movement possible. Without joints, the human body would be stiff and unbending. Think of every place in the body where movement can occur. There is a joint there, allowing that movement to happen.

There are several different types of joints in the human body. Each type has its own special function or role. Some joints allow for lots of movement, while others move just a little or not at all.

Movable joints are surrounded by a joint capsule. These capsules have a *synovial* (suh-NOH-vee-ul) *membrane,* which produces *synovial fluid.* Synovial fluid is a watery liquid that lubricates the joint area, allowing for ease of movement.

The most common type of joint in the human body is the *hinge joint.* These joints enable back and forth movement, just like the hinges of a door. Hinge joints in the elbows, for example, allow humans to flex and straighten their arms. Hinge joints are also found in the knees, fingers, and toes.

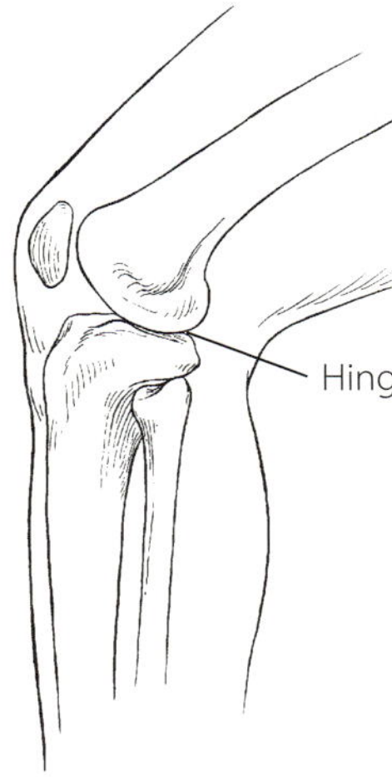

Ball-and-socket joints allow the greatest range of movement. Bones connected by this type of joint can move in almost any direction. Ball-and-socket joints can be found in the hips and shoulders. The ball-and-socket joints of the shoulders enable humans to move their arms in complete circles.

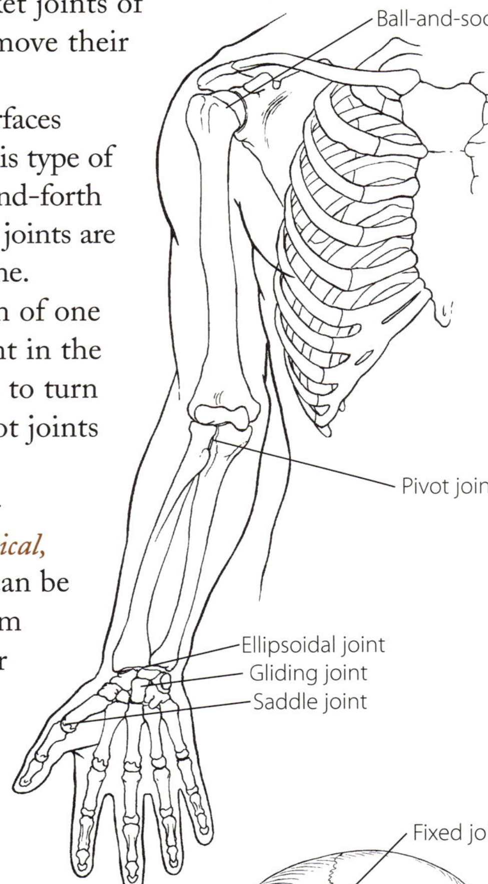

Gliding joints have two flat surfaces that slide easily over each other. This type of joint allows for very limited back-and-forth or side-to-side movement. Gliding joints are found in the ankles, wrists, and spine.

Pivot joints allow the rotation of one bone around another. A pivot joint in the neck, for example, allows humans to turn their heads from side to side. Pivot joints are also found in the lower arms.

In an *ellipsoidal joint,* an egg-shaped bone end fits into an *elliptical,* or oval, cavity. This type of joint can be extended and flexed or moved from side to side. However, it allows for only limited rotation. This type of joint is found in the wrists.

Found only at the base of the thumbs, *saddle joints* allow bones to rock back and forth and side to side. This joint gets its name from its shape—it resembles a saddle.

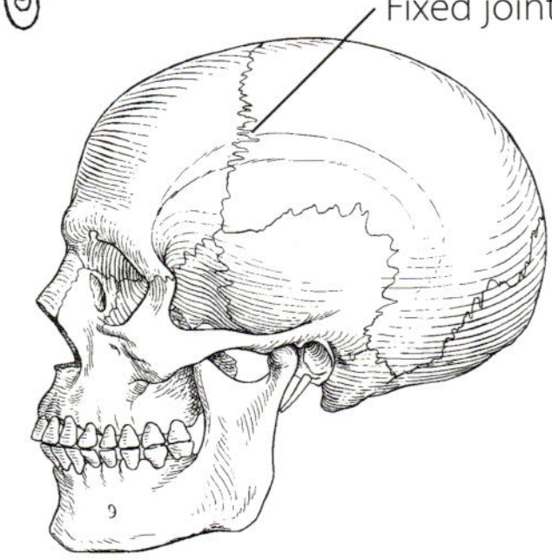

Fixed joints (also known as sutures) allow virtually no movement at all. Bones connected by fixed joints are very close together. This type of joint is found in the skull and the pelvis.

Fascinating Joint Facts

- The knee joint is the largest joint in the human body.
- The hip joint is the strongest joint in the body. It must support the full weight of the upper body.
- A "double-jointed" person doesn't really have extra joints. Instead, the ligaments in the person's body can stretch farther than other people's. This allows bones to move more than normal.
- Each joint in the body has its own name. For example, the joint where the bones of the upper and lower arm meet is the elbow.

CONNECTIVE TISSUES

Bones are connected to other bones and muscles by different types of connective tissue. Some types of connective tissues are strong, flexible fibers that work with muscles and bones to make movement possible. The substances that make this type of connective tissue stretchy and flexible are the proteins collagen and *elastin.*

There are three main types of tissue that connect bones to other bones and muscles: ligaments, tendons, and cartilage. *Ligaments* are strong tissues that connect bones to other bones across a joint. Ligaments are more stretchy than cartilage and tendons. They reduce friction between bones.

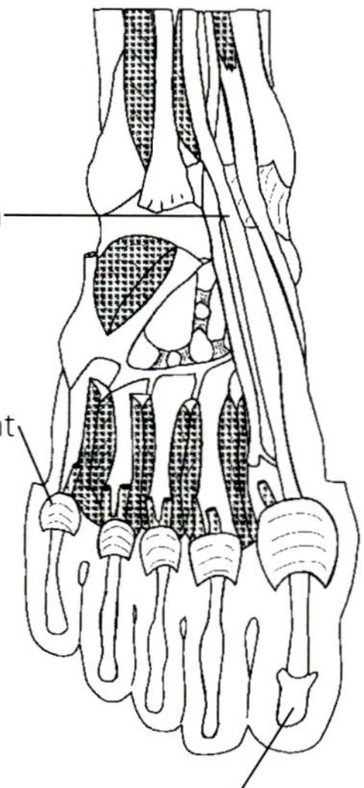

Tendons are the strong, elastic fiber cords that connect muscles to bones or other muscles. *Bursae* (BUR-see), capsules filled with synovial fluid, lubricate the spots where tendons meet bones. Like the other types of connective tissue, tendons prevent friction problems.

Cartilage (KART-ih-lij) is a strong, flexible connective tissue that covers the bone ends. Cartilage protects bones from damage by absorbing the shock of everyday activities. It prevents the bone ends from being worn away by the constant friction on them. Cartilage also gives shape and flexibility to the body. For example, cartilage at the

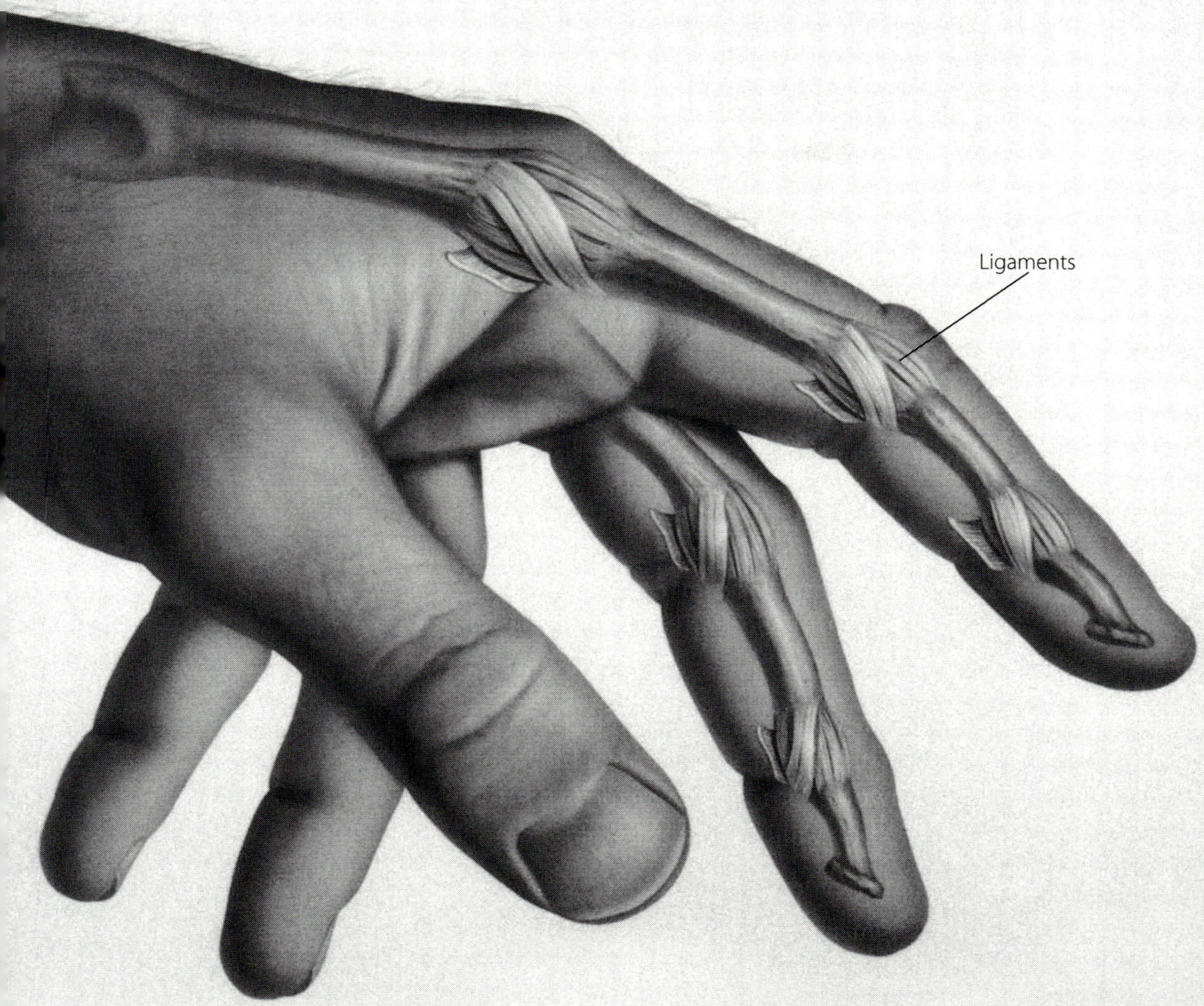

Snap, Crackle, Pop

What causes the popping sound when people crack their knuckles and other joints? The synovial fluid of movable joints contains dissolved oxygen, carbon dioxide, and nitrogen gases. When a joint is moved in a certain way, air and gas pop out of the fluid. The more gas that is released, the louder the noise. The joint cannot be cracked again until the gas bubble dissolves back into the synovial fluid. Cracking and popping may also be heard when ligaments and tendons move across bumps in bones.

end of the nose and ears adds shape and flexibility to those parts of the body.

There are several different types of cartilage. *Hyaline* (HYE-uh-lin) cartilage is the most common type of cartilage in the human body. This cartilage covers the bone ends in movable joints. It also covers the ends of ribs where they join the *sternum,* or breastbone. Most bones start out as hyaline cartilage before they harden into true bone.

As its name implies, *elastic cartilage* is elastic and allows vibration and movement. Elastic cartilage is found in the ear and the epiglottis. The *epiglottis* (eh-pih-GLAHT-tiss) is the flap that closes over the *larynx* (LAAR-inks), or vocal cords, when humans swallow to prevent food from going down the wrong way.

Fibrocartilage (FYE-broh-kart-ih-lij) is also elastic, yet very strong. This type of cartilage is found between the vertebrae. It provides support for the spine and allows the spine to twist and bend. Fibrocartilage is also found in the pubic area.

2

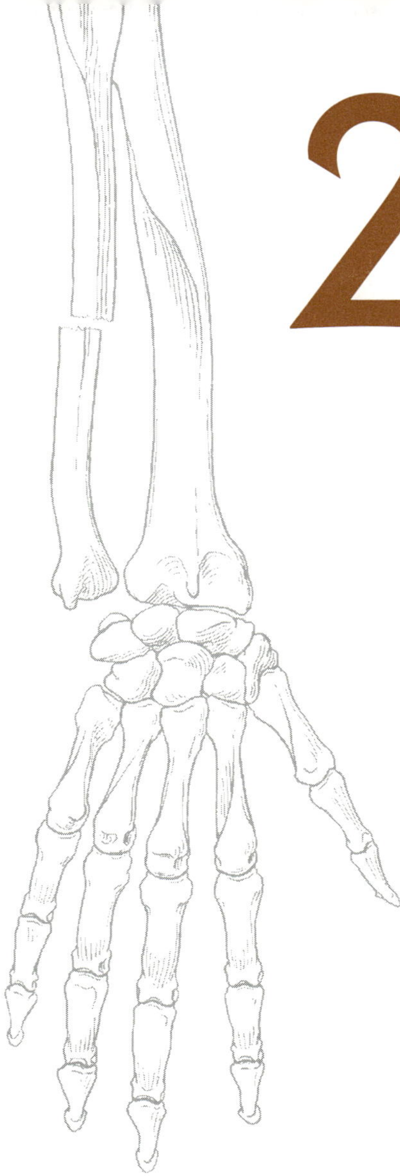

BONE DISORDERS

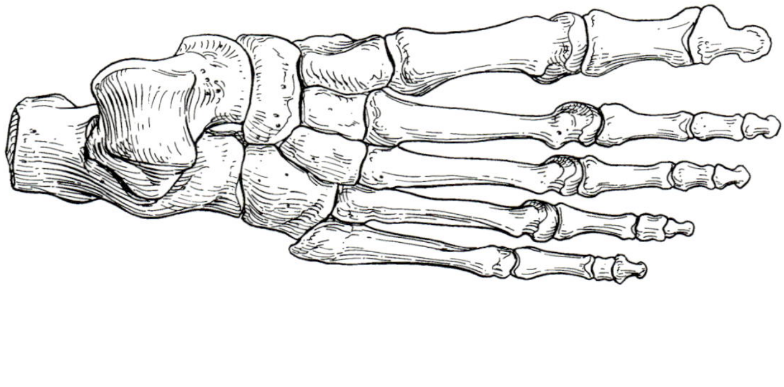

A number of health conditions that can affect the body's skeletal system might require the attention and care of a medical professional. These conditions include injuries, disease, and inherited disorders.

FRACTURES

Each and every day, you use your bones. You use them when you stand, work, play, eat, or stretch. Even when you're resting, your bones are supporting your weight.

Although bones are strong and flexible enough to withstand your daily routine, they can still be injured and broken. A break in a bone is called a *fracture.* A fracture can be caused by a sudden trauma, such as an accident or fall. A fracture might also be caused by repetitive or constant force or pressure upon a bone. This type of fracture is known as a *stress fracture.* Certain conditions or diseases, including anorexia and osteoporosis, can make bones more prone to fractures.

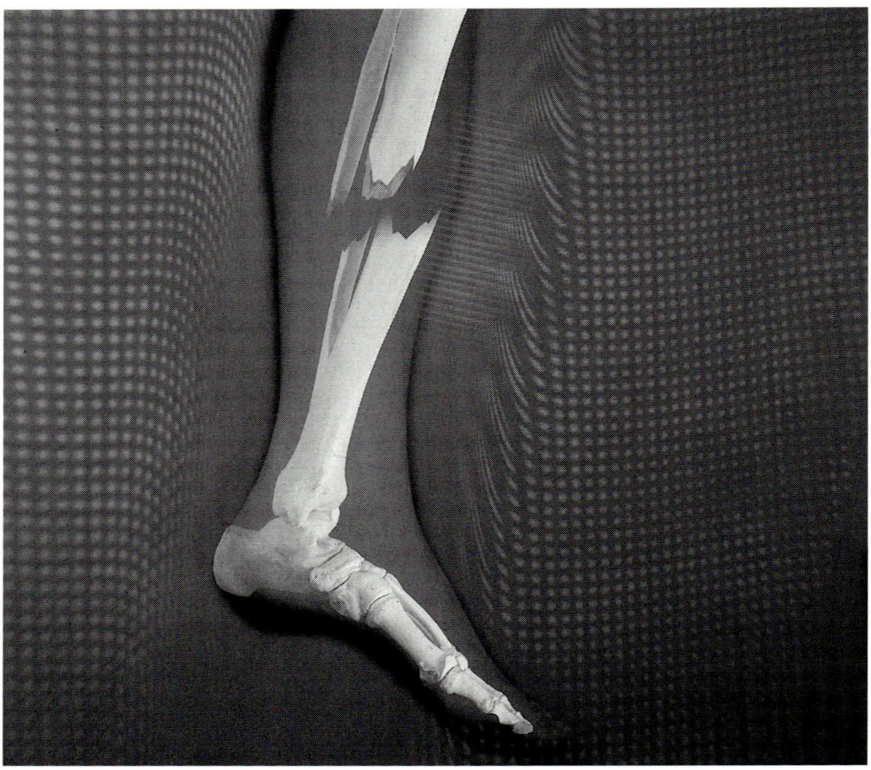

The lower leg with a simple fracture of the tibia (the large, inner bone of the lower leg) and fibula (the outer and smaller of the two bones of the lower leg)

Fractures can happen to anyone. Some bones, such as the forearm bones, shinbones, and collarbones, are more frequently fractured than others. Each year, more than 6.5 million Americans, young and old alike, suffer fractures. In fact, each American will break an average of two bones in his or her lifetime.

As people grow older, the causes of bone fractures change. Children often break bones while playing or taking part in other physical activities. People between the ages of twenty and fifty often break bones in car accidents or work-related accidents. The elderly, particularly those with osteoporosis, may break bones while doing even the simplest of tasks.

Other people's lifestyles put them at increased risks for bone fractures. Athletes, for example, are often prone to stress fractures as a result of performing the same actions over and over. The most common types of stress fractures in athletes occur in the feet, ankles, and shins. Running and other activities put constant pressure on these parts of the body.

There are several different types of fractures. Some are more serious than others. The seriousness of a fracture depends upon the age of the injured person, any diseases that may make the fracture worse, and the force involved in the injury.

A *greenstick fracture* is a break that does not go all the way through the bone. This type of fracture is most common in children. Because their growing bones are not completely hardened, they are more springy and flexible than adult bones. Greenstick fractures are usually quick to heal. Doctors often immobilize the affected bone with a cast until the break has healed.

In a *bending fracture,* a bone is bent, but does not break. This type of fracture occurs only in children.

A *simple fracture,* also known as a close fracture, occurs when the bone is broken but does not break through the skin. Simple fractures are often treated with casts.

A *compound fracture,* also called an open fracture, is much more serious than a simple fracture. Here, the bone is completely broken and the skin is pierced by the bone ends. Sometimes, screws must be inserted into the bone to hold the broken pieces together. With this type of fracture, there is also the chance for *osteomyelitis,* or bone infection.

A *comminuted* (KAHM-ih-noot-ed) *fracture* involves bone that is crushed, shattered, or splintered into several pieces. This type of fracture is often a result of motor-vehicle accidents. However, people with osteoporosis may suffer a comminuted fracture after a fall or a minor injury. Surgical methods may be needed to reconstruct the bones after this type of injury.

A young boy sits on the sidelines with his fractured leg in a plaster cast. The cast immobilizes the fracture so it can heal properly.

FIXING A FRACTURE

A person who suffers a fracture may hear an audible crack as the bone breaks. Symptoms of a fracture include pain, tenderness, and swelling of the affected area. A person who has broken a bone may not be able to move the injured body part. In some cases, the bone may poke through the skin. People who think that they have fractured a bone should seek medical attention immediately.

When bone is injured, it immediately begins to repair itself. Blood from the fractured bone pieces forms a blood clot, which surrounds the broken sections. Osteoblasts move into the injured area and begin laying down new tissue to replace the damaged tissue.

Eventually, the cells create a bridge of collagen, called a *callus,* between the fractured parts. The callus will harden and form new bone. New bone is just as tough and strong as the bone that it replaces.

Because the bones do all this repair work themselves, many fractures just need to be left alone to heal. Doctors help along the natural healing process by immobilizing the fractured bone and surrounding joints. This means making sure that the bone is held securely in the proper position until healing is completed. Doctors may use a plaster or fiberglass cast to set the bone. Other means of immobilization include splints, slings, and braces.

A badly fractured bone may require surgery to reposition and immobilize it. In some cases, pins may be inserted through the skin and into the fractured bone. These pins are connected to metal frames outside the body that hold the bones in place. The pins are removed once the bone has healed. Doctors may repair more serious fractures by inserting plates, screws, pins, or nails directly into the bone. The devices are surgically removed once the bone has healed.

Fast Fact

Fractured bones in children can take from six to eight weeks to heal. For adults and older children whose bones have stopped growing, a fracture may take between ten and twelve weeks to heal. Severe breaks may take even longer.

An Electric Treatment

Sometimes, doctors use electricity to heal long bone fractures. Medical experts say that electrical charges promote bone growth and help fractures heal more quickly. Studies suggest that with electricity, fractured bones may heal in half the time. Different types of devices are used to give electrical treatment. Some devices are designed to be used on the outside of the body. They are placed on the fractured area and emit electrical charges through the skin to the bone. Other tiny, electricity-generating devices are surgically implanted within the body to direct electrical charges close to the bone.

Sometimes, the ends of two bones may become disconnected from each other. This is known as a *dislocation.* A dislocation most often occurs as a result of an injury or accident. Ligaments and tendons may be torn in the process, and nerves can be damaged. The most frequently dislocated bones are the jaw, shoulders, knees, elbows, and fingers. Symptoms of a dislocation include pain, swelling, and difficulty moving the affected area. Doctors treat a dislocation by moving the bone back into its proper place.

Skateboard and Scooter Injuries

The American Academy of Pediatrics (AAP) reports that the number of injuries due to skateboards and scooters continues to rise. According to the Consumer Product Safety Commission, 104,000 people were treated for skateboard-related injuries in the emergency room in 2001. For January 2001 through September 2001, 84,400 people were treated in emergency rooms for scooter-related injuries. The three most commonly injured areas were the ankle, wrist, and face. With people riding scooters, one-third of all injuries were bone fractures or dislocations. To avoid injury, experts recommend that children take the proper safety precautions when riding scooters. Such precautions include wearing a helmet and other protective gear, watching for debris or obstacles in pathways, and riding safely and responsibly.

COMMON FRACTURES

Colles's fracture is the most common type of break in people under the age of seventy-five. This injury to the bones in the lower arm usually occurs when a person holds out a hand to break a fall. It is also called a buckle fracture, because the impact of the ground on the wrist causes the bones in the forearm to buckle. Children frequently suffer Colles's fractures.

The *tibia,* or shinbone, is often fractured by young people during sports or other physical activity. This bone bears much of the weight of the legs, especially when running or exercising. This type of break is sometimes accompanied by a fractured *fibula,* the other long bone in the lower leg.

The *clavicle,* or collarbone, is another commonly broken bone. The clavicle connects the breastbone to the *scapula,* or shoulder bone. All pressure placed on the shoulder is transferred to the clavicle. As a result, any severe pressure or force on the shoulder can result in a broken clavicle.

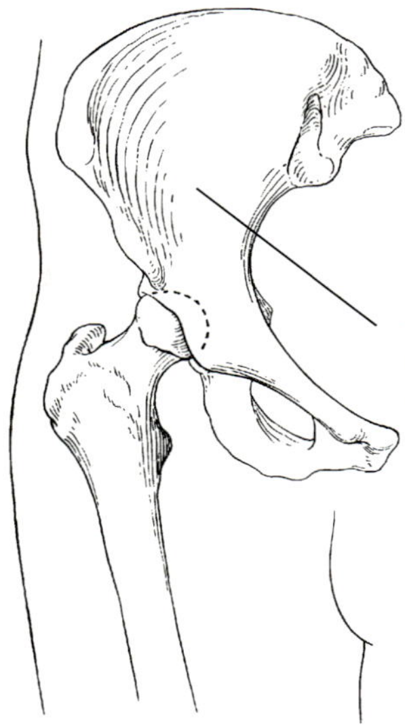

A fractured *hip* is common in elderly people. Ninety percent of all people hospitalized with fractured hips are over the age of sixty-five. As people age, their bones lose density, becoming fragile and easily breakable. A fall—even a minor one—can often result in a fractured hip or other injury for an elderly person. Hip fractures can be especially serious. The American Academy of Orthopedics says that 24 percent of hip fracture patients over the age of fifty die within twelve months of suffering their injury due to complications resulting from the injury.

Spinal fractures are serious injuries that result from the severe compression, abnormal rotation, or bending of the spine. With this type of fracture, it is important that the spinal cord be completely immobilized. If not, permanent damage could result, including partial or complete paralysis. *Paralysis* (puh-RAL-ih-siss) is the loss of movement or sensation in part of the body.

GROWTH PLATE INJURIES

Growth plates are areas of tissue at the ends of long bones in children and adolescents. Because the growth plate has not yet been replaced by solid bone, it is the weakest section of the bone. When a fracture occurs on a growth plate, long-lasting damage can result if the injury is not treated properly. A damaged growth plate can permanently affect the growth and development of the bone. Most growth plate injuries are the result of a fall or other traumatic injury. These types of injuries can also be caused by overuse. Pitchers, gymnasts, and long-distance runners are sometimes affected by growth-plate injuries. Frostbite can also damage children's growth plates. Common growth-plate injuries occur in the wrist, leg, ankle, foot, and hip.

Once doctors diagnose a growth-plate injury, they usually immobilize the area by placing a cast or splint on it. Surgery may be required to get the bones back to their proper positions. Over the next two years, doctors will probably X-ray and watch the bone to make sure that it is developing and growing properly. About 85 percent of all such injuries result in full and complete recovery.

Little League Elbow

As more and more kids join youth baseball teams, sports-related injuries are becoming more common. The CDC reports that each year, 125,000 baseball and softball players under the age of fifteen seek emergency-room treatment for sports-related injuries.

Pitchers are especially at risk for an injury nicknamed "Little League elbow." Little League elbow is an overuse injury caused by repetitive throwing. It occurs in kids whose bones are still growing. The repetitive, odd movements of pitching cause bones to chip and growing areas of bone to become inflamed. The condition may also cause problems with the tendons and joints of the arm.

The CDC estimates that as many as 45 percent of all pitchers under the age of twelve suffer from this type of injury. For high school pitchers, that figure rises to as high as 80 percent. Kids showing signs of Little League elbow should stop pitching immediately and rest until they have recovered completely. It may take months before the elbow has healed. In some cases, surgery may also be required to repair the damage.

Ignoring the injury can lead to permanent damage, including deformity, constant pain, and arthritis later in life. Experts say that the best way to avoid such problems is to take preventive action. Here are some pitching tips from the experts that will help kids stay in top shape.

- Pitch no more than six innings per week or throw no more than 90 to 100 pitches per outing.
- Take at least three days off between games.
- Pay attention to any elbow pain and report it immediately. Stop the offending activity and see a doctor. *Never* play through the pain or just shake it off.
- Avoid throwing curveballs and sliders. These types of pitches put extra strain on the elbow, especially if the pitch is not thrown correctly.
- Excessive practice can result in injury, not perfection.
- Learn to throw properly—not fast, not hard, but correctly. Proper technique will minimize injuries.

Opposite:
The repetition and unusual motion of pitching a baseball can damage bones and ligaments of the arm, especially near the elbow and wrist.

OSTEOPOROSIS

Osteoporosis (ah-stee-oh-puh-ROH-siss) is a disorder that causes the loss of the minerals that make bone dense and strong. As people age, their bones become thinner and more fragile, which makes them much more susceptible to fractures. Back, hip, and wrist fractures are common injuries in people with osteoporosis.

Osteoporosis is the most common type of bone disease in the United States. Although it is most often found in older people, it can affect people of all ages. According to the National Institute of Arthritis and Musculoskeletal and Skin Diseases (NIAMS), 10 million people in the United States suffer from osteoporosis. The group also reports that 18 million more Americans have low bone density, a condition that places them at risk for osteoporosis.

Osteoporosis is much more prevalent in women than in men. American women are four times more likely to develop the condition than are American men. NIAMS estimates that half of all women over the age of fifty will suffer an osteoporosis-related bone fracture at some point in their life.

Fast Fact

The word *osteoporosis* comes from the Greek words *osteon,* meaning "bone," and *poros,* meaning "passage" or "space."

Why are women at higher risk than men? As women age and go through menopause, they lose hormones, such as *estrogen* (ESS-troh-jen), that help bones absorb calcium and properly use calcium. At *menopause,* women stop menstruating and their bodies produce fewer of such important hormones as estrogen. After menopause, women can lose as much as 25 percent of their bone mass. This loss may lead to osteoporosis.

What causes osteoporosis? Some people have a family history of osteoporosis. White and Asian women who are thin, small, and fair skinned are also predisposed to the condition. In addition, a number of health conditions and behaviors can increase the risk of osteoporosis. People with anorexia nervosa, for example, have an increased risk of getting this disease. *Anorexia nervosa* (an-uh-REKS-ee-uh ner-VOH-suh) is an eating disorder. In addition, alcohol consumption, cigarette smoking, and steroid use can add to the risk of getting osteoporosis. Some researchers believe that environmental factors also increase the risk of osteoporosis. For example, lead pollution and ingestion may contribute to this bone disorder.

TREATING OSTEOPOROSIS

Osteoporosis is normally diagnosed by a *bone mineral density* (BMD) *test*. This noninvasive test is easy and painless and can be performed wherever a machine is available. Some machines measure the density of bone in the spine or hip. Portable machines can measure bone density of the finger or ankle. Some BMD tests use X-ray machines or CT machines to check the body for bone density. Unfortunately, the condition is often not discovered until a person breaks a bone when lifting a heavy object or suffering a minor fall.

Osteoporosis can be prevented, especially when people take steps early in life. Children and young adults can strengthen their bones by eating a healthy, calcium-rich diet. The proper intake of calcium and vitamin D is one of the most effective ways of avoiding osteoporosis. Older people can take mineral supplements to get enough calcium. In addition, exercise is a benefit for people of all ages. It helps keep bones strong and healthy.

People with osteoporosis are treated in a number of ways. Hormones, sodium fluoride, and a number of drugs can be used to stimulate bone growth and boost bone density. People with osteoporosis should also take special care to avoid damage. They should maintain good posture, wear sensible shoes, and adjust their living space to minimize the chances of a fall.

Researchers are currently studying osteoporosis to determine what can be done to decrease the incidence of this condition. They are studying bone mass and calcium intake in children. They are also researching and identifying genetic factors—those that have been inherited from previous generations—and other reasons that may make people likely to develop this debilitating bone disease. In addition, scientists are studying how osteoporosis affects different populations of people.

Calcium Crisis

A 2002 study by the National Institutes of Health (NIH) showed that half of all children under the age of five don't get enough calcium in their diets. The same study revealed that 85 percent of teen girls and 60 percent of teen boys don't get the recommended daily amounts of calcium, either. The NIH calls this a "calcium crisis."

Osteoporosis Risk Factors

Here are a few of the factors that increase a person's chances for getting osteoporosis.

- increasing age
- female gender
- white or Asian race
- early menopause
- thin, small-boned body
- family history of osteoporosis
- fair skin, blue eyes
- few or no children
- low calcium intake
- lack of exercise
- cigarette smoking
- alcohol consumption
- high caffeine consumption
- high-protein diet
- high-salt diet
- lack of mobility
- certain other diseases or conditions, including hormone disorders, anorexia nervosa, jaundice, malnutrition, carcinoma, diabetes, and rheumatoid arthritis
- use of certain drugs, including some steroids

BONE CANCERS

Cancer of the bones can affect people of any age, race, or sex. According to the American Cancer Society, about 1,400 Americans died from cancers of the bones and joints in 2001. Researchers are not sure what exactly causes bone cancer. They do know that genetic traits, radiation and chemotherapy, and certain diseases can increase the risks of developing bone cancer.

People with bone cancer often experience pain, loss of flexibility and bone strength, and swelling near the affected area. Other symptoms include fever, fatigue, weight loss, and *anemia* (uh-NEE-mee-uh), a reduction in the number of red blood cells in the body. Doctors use blood tests, X-rays, bone scans, and other means to help them diagnose bone cancer. The final diagnosis must be made with a bone *biopsy*. A biopsy (BYE-ahp-see) is a surgical procedure to remove and examine tissue from the tumor itself.

BONE TUMORS

Bone tumors are often the first indication of bone cancer. A *tumor* is a painful swelling on the bone. However, not all bone tumors are *malignant,* or cancerous. Some tumors are *benign,* or noncancerous. Benign tumors, unlike malignant ones, do not spread and are not life-threatening. The cause of benign bone tumors is unclear. They may be the result of rapid bone growth, a traumatic injury, or a hereditary trait. A doctor will keep an eye on a benign tumor by X-raying it periodically. Some benign bone tumors disappear without any treatment. In other cases, tumors must be removed surgically. Benign bone tumors include *osteochondromas* (ah-stee-oh-kahn-DROH-muhz), *osteomas* (ah-stee-OH-muhz), and *osteoblastomas* (ah-stee-oh-blass-TOH-muhz).

TYPES OF BONE CANCER

There are two main types of bone cancers: primary and secondary. *Primary bone cancer* starts in the bones. This type of cancer is much rarer than secondary bone cancer. It occurs most frequently in young people.

Osteosarcoma (ah-stee-oh-sar-KOH-muh) is the most common type of primary bone cancer. It affects the long bones of the body, especially the femurs. Osteosarcoma, which starts in the new tissue of growing bones, most often affects people between the ages of ten and thirty.

Chondrosarcoma (kahn-droh-sar-KOH-muh) affects the pelvis, ribs, and breastbone. It starts in cartilage tissue. Chondrosarcoma rarely affects people under the age of twenty.

Ewing's sarcoma (sar-KOH-muh), also known as Ewing's tumor, occurs most frequently in children. This type of cancer begins in the bone marrow. Long bones are most often affected.

Fibrosarcoma (fye-broh-sar-KOH-muh) most frequently occurs in older people. It most often affects the jaw, leg, and arm bones.

A *chordoma* (kor-DOH-muh) is a cancerous tumor that appears at the base of the skull or on the spine. This type of tumor may recur, even years after successful treatment.

Secondary bone cancer starts in another part of the body and then spreads to the bones. It is much more common than primary bone cancer. It is also called *metastatic,* or spreading, bone cancer. Secondary bone cancer often occurs in older people. Areas of the body that are commonly affected by this type of cancer include the skull, sternum, pelvis, vertebrae, and ribs.

Some types of cancer start in the bones but are not considered bone cancers by doctors. One such cancer is *myeloma* (mye-uh-LOH-muh). Myeloma begins in plasma cells inside bone marrow. *Plasma* is the clear liquid that makes up more than half of all blood. Like other cancerous bone tumors, myeloma can cause destruction of the bone. Another type of cancer that sometimes starts in bone marrow is *non-Hodgkin's lymphoma* (lim-FOH-muh). Non-Hodgkin's lymphoma may start in *lymphocytes* (LIM-foh-sites), white blood cells that fight bacteria and infection.

TREATING BONE CANCER

Bone cancer, like other forms of cancer, is a serious medical condition that can be life threatening if not caught early and treated effectively. The type of cancer treatment that each patient receives depends upon the type of cancer he or she has, as well as other factors. The doctor's goal in treating bone cancer is to remove the cancerous tumor, prevent it from spreading, and prevent a recurrence.

Bone grafting is one type of surgery that is sometimes used to treat primary bone cancer. Cancerous bone is removed and new, healthy bone is grafted, or transplanted, in its place. In some cases, limbs affected by cancer may have to be *amputated,* or surgically removed.

Radiation treatment is the use of X-rays and other radiation sources to kill cancerous cells. Used on both primary and secondary bone cancers, radiation treatment can reduce the size of a tumor. Doctors can then use anticancer drugs to treat the malignant areas. In some cases, the cancer can be completely removed through this type of treatment.

Chemotherapy (kee-moh-THERR-uh-pee) is the use of anticancer drugs to cure or stop the spread of cancers. Because this type of treatment affects the entire body, chemotherapy is most likely to be used when a cancer has spread throughout the body or is found in more than one place. It is also used when cancer cannot be treated surgically. Chemotherapy is often used with other cancer treatments.

Long-term chemotherapy treatment can cause serious health problems. It damages the bone marrow and some cells, resulting in anemia and a higher risk of infections. Long-term effects of chemotherapy include damage to the heart or kidneys. Patients undergoing chemotherapy often lose their hair and may experience severe nausea and vomiting.

BIRTH DEFECTS AND GENETIC DISORDERS

A *birth defect* is any physical problem that an infant has at birth. Researchers have identified more than 4,000 different types of birth defects. Each year, 3 out of every 100 babies born in the United States are born with a serious birth defect.

Birth defects can affect any part of the body, including the musculoskeletal system. Two musculoskeletal birth defects are *cleft lip* and *cleft palate.* Together, these two conditions are the fourth most common type of birth defect in the United States. As many as one out of every 700 babies is born with a cleft lip, cleft palate, or both.

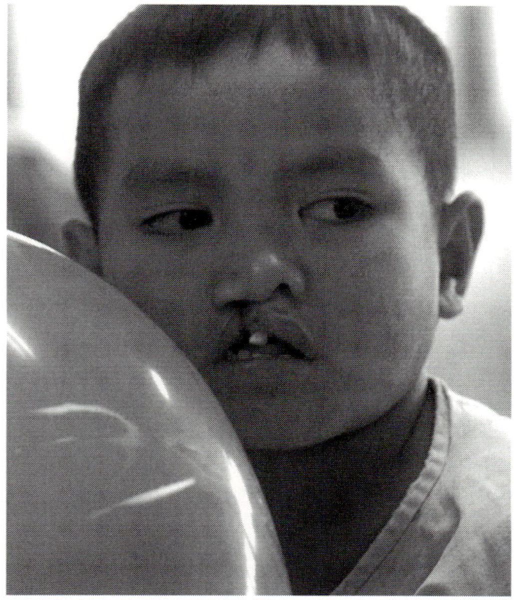

This young boy is awaiting surgery to repair a cleft palate.

Cleft lip is a permanent split in the upper lip. This split may extend to include the bones of the upper jaw and upper gum. Cleft palate is a genetic condition in which bones in the roof of the mouth do not fuse. Both conditions occur early in a woman's pregnancy.

Infants born with these birth defects have a high risk of ear infections because the Eustachian tubes in the ears do not drain properly, causing fluid collection, ear pressure, and infection. Babies with clefts are also at increased risk of colds, speech defects, and tooth problems. Doctors often recommend surgery, dental work, and speech therapy to correct cleft conditions.

Other birth defects are *Crouzon syndrome,* Treacher-Collins syndrome, clubfoot, and Klippel-Feil syndrome. Crouzon (kroo-ZAHN) syndrome, also known as craniofacial dysostosis (kray-nee-oh-FAY-shul dis-oss-TOH-siss), is an abnormal fusing of some of the bones in the face and skull. One of the chief problems caused by Crouzon syndrome is underdevelopment of the upper jaw. This can cause bulging eyes and give the middle part of the face a sunken appearance. In addition, infants born with this condition are at risk for ear problems, hearing loss, and abnormal speech development.

Treacher-Collins syndrome can affect the size and shape of the eyelids, ears, cheekbones, and jaws. As with other types of facial disorders, the hearing can be affected. This disorder can also affect breathing and feeding of infants.

Clubfoot is a condition in which the bones, joints, and muscles of the foot are deformed: the foot is stiff and turned inward. Children born with clubfoot are treated with casts and splints that help correct the shape of the foot. Surgery may be needed to correct bone and joint placement. Even after surgery, however, problems may remain. Later in life, those born with clubfoot may suffer pain. To ease the pain, they may need to wear special shoes or even have more surgery.

With *Klippel-Feil* (klip-ul-FYLE) syndrome, a child is born with two of the vertebrae in the neck fused together. This condition occurs early in a pregnancy. It results in a short neck and restricted mobility of the upper spine. Klippel-Feil often accompanies other genetic conditions, including scoliosis, spina bifida, and cleft palate. Surgery may be required to correct physical deformities. Physical therapy is also usually recommended.

Genetic disorders are abnormalities in a person's genes or chromosomes. These defects may be inherited from a parent, or they may develop on their own. Genetic disorders may be immediately apparent when a child is born or may be invisible at first, possibly causing problems years later.

Genetic disorders that affect the musculoskeletal system include *osteogenesis imperfecta,* Pierre Robin sequence, and Apert syndrome. Osteogenesis imperfecta (ah-stee-oh-JEN-ih-siss im-per-FEKT-uh) is an inherited condition that affects protein production in the body. Bones are easily breakable and sometimes deformed. In addition, some people who have this condition are much shorter than usual, with loose joints and poor muscle development.

Pierre Robin (pee-AYR roh-BAN) sequence is a disorder in which the lower jaw of a fetus does not develop properly. As a result, the tongue is positioned more to the back of the mouth. Here, it can obstruct the airway and cause feeding and breathing problems. If breathing problems are severe enough, surgery may be necessary. This condition can also be caused by fetal alcohol syndrome, an environmental disease. Children may outgrow the condition by the age of six if their lower jaws grow rapidly enough.

In *Apert* (ay-PAYR) *syndrome*—thought to be genetically inherited—the seams between a baby's skull bones close too early. This gives the head a peaked look and affects the structure of the face, as well. In addition, the soft tissue and sometimes the bones of the fingers and toes may fuse together as the child matures. A child with Apert syndrome may also be short in stature.

MISCELLANEOUS CONDITIONS

Doctors are unsure exactly what causes some skeletal problems. Conditions with unknown causes include three different types of curvature of the spine.

SCOLIOSIS

Scoliosis (skoh-lee-OH-siss) is a sideways curvature of the spine, usually in an S shape or a C shape. It occurs most often in children and adolescents, but older people may also suffer from spinal curvatures. About 2 percent of the U.S. population shows some degree of scoliosis.

In eight out of ten cases, the exact cause of scoliosis is unknown. Because the condition does seem to run in some families, researchers believe that it may be caused by a genetic factor. In older people, spinal curvatures may be caused by osteoporosis. Other cases may be related to poor posture or weak abdominal muscles.

To treat scoliosis, doctors carefully watch how the curve is developing. They take periodic X-rays to follow the curve's progress. In some cases, no treatment at all may be needed, especially in children. Children's spines may straighten naturally as they age. If doctors feel that treatment is necessary, they may use a back brace to prevent the curve from worsening. In severe cases, surgery may be required to stabilize the curve.

KYPHOSIS

Also known as Scheuermann disease and roundback, *kyphosis* (kye-FOH-siss) is a bulging of the upper part of the spine. It is sometimes associated with scoliosis. In addition to a rounded upper spine, pain and stiffness are characteristic of this condition.

Kyphosis can cause lung and heart problems as the spine squeezes these organs. This disorder may result from a trauma, developmental problems, or degenerative problems. When it occurs in adolescents, the exact cause is generally unknown. In the elderly, kyphosis can be caused by osteoporosis.

LORDOSIS

Lordosis (lor-DOH-siss), also known as swayback, is an inward curve of the spine just above the buttocks. Usually, if the back is still flexible and easily moved, doctors will not treat the condition.

PAGET'S DISEASE

Paget's (PAA-jitz) disease is another bone disorder that has an unknown cause. Also called osteitis deformans, Paget's is a condition that results in large, deformed bones. Paget's disease affects mainly older people and has a higher incidence in the United Kingdom than any other place in the world. In Paget's disease, bone cells begin rapidly breaking down bone and replacing it with abnormal bone. The new bone expands, disfiguring a person. The new bone is less dense and weaker than normal bone and more prone to fractures.

Paget's disease often affects the skull, spine, pelvis, and leg bones. It can cause an enlarged skull, arthritis, bowed legs and spine, loss of hearing, heart disease, and many other conditions.

PERTHES DISEASE

Perthes (PURTHS) *disease* usually occurs in children aged three to eleven, particularly boys. Doctors do not know what causes the condition. Perthes is marked by abnormal blood circulation in the femur. Because of this, the growth plate at the head of the femur becomes soft and sometimes flattens out. The condition causes pain in the groin area, along with a limp.

Children with this condition are usually treated with plenty of rest, as well as casts or braces that must be worn until the bone has healed. People suffering from Perthes usually recover completely. However, severe cases may require surgery to repair the damaged bone and may result in chronic, or long-lasting, stiffness and even arthritis of the hip.

3

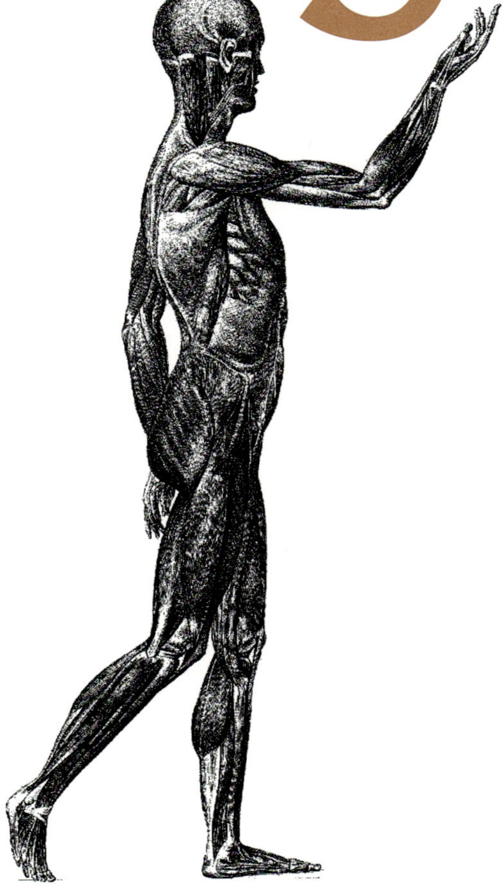

MUSCLE DISORDERS

Like other parts of the body, the muscles can be damaged by injury or disease. Here are a few of the disorders that can affect the body's many muscles.

MUSCLE INJURIES

Most of the problems that affect the muscles are a result of injuries rather than disease. The most common muscle injuries include cramps, strains, tears, and contusions.

Cramps are painful, involuntary muscle contractions. They may occur when a person is tired or dehydrated, has overexercised, or lacks oxygen or other nutrients that keep the muscles working properly. Cramps may also be caused by an accumulation of waste products, such as lactic acid, in the muscles. People who experience cramps when swimming or running often have not warmed up and stretched their muscles enough before the exercise.

Cramps are temporary, eventually fading away as the muscles relax. Usually, the only medical treatment that is needed is slow stretching of the affected muscle. Sometimes, cramps seem to last forever. If cramps occur frequently and last a long time, the person experiencing them should talk to a health professional.

Spasms, like cramps, are involuntary muscle contractions. However, spasms may happen in a split second and then be over. A tiny twitch in your eye, for example, is a spasm.

Montreal Canadiens hockey player Martin Rucinsky lies on the ice after receiving a painful muscle contusion in a collision with another player.

Strains and *tears* are much more serious than cramps. When a muscle is moderately damaged, it is said to be strained. A more severe muscle injury is a tear. When a muscle is torn, a person has pain and swelling.

Another type of muscle injury is a *contusion* (kun-TOO-zhin). This is a blow or hit to the muscle that causes injury. A contusion may cause bleeding of the muscle, which may result in a bruise on the skin and underlying tissue. A contusion might also result in buildup of calcium deposits in the muscle tissue. This can cause even more pain and stiffness in the injured area. In severe cases, surgery to remove the deposits may be necessary.

MUSCULAR DYSTROPHY

Genetic disorders may be inherited from a parent, or they may develop on their own. Genetic disorders can affect any part of the human body. One type of genetic disorder that affects the body's muscles is *muscular dystrophy* (DISS-truh-fee), or MD. Muscular dystrophies are a group of diseases in which the skeletal muscles weaken and deteriorate. Muscles gradually waste away until they are completely unusable. With some types of muscular dystrophy, the cardiac muscles and other smooth muscles are also affected. Certain organs may be affected, as well.

MD is usually inherited. People who develop the disease are generally born with a genetic abnormality that causes skeletal muscles to develop improperly. People with MD may eventually become physically disabled and have difficulty walking, breathing, or even sitting upright.

MD affects people of all ages. Some infants and children are stricken with the disease, while other people don't get MD until they are in middle age. Some forms of the disease progress slowly, and patients can lead a normal life for many years. Other forms progress more rapidly, resulting in severe disability and a shortened life span.

Doctors use blood tests, genetic tests, and muscle biopsies to diagnose MD. These tests help doctors know how to best treat a patient. Doctors treat some types of MD with physical therapy, exercise, and medicines. Surgery to correct such symptoms as curvature of the spine may also be possible for some patients.

There is currently no cure for MD. Scientists and doctors continue to research the condition, looking for causes, better treatments, and a cure to end this debilitating disease.

Health and History: Understanding Muscular Dystrophy

Today, doctors know that muscular dystrophies are inherited disorders. In the past, this wasn't the case. Doctors, watching their patients' muscles wasting away, believed that the condition was caused by a problem with the diet. As a result, the condition was named muscular dystrophy. Dystrophy comes from Latin and Greek words meaning "faulty nutrition."

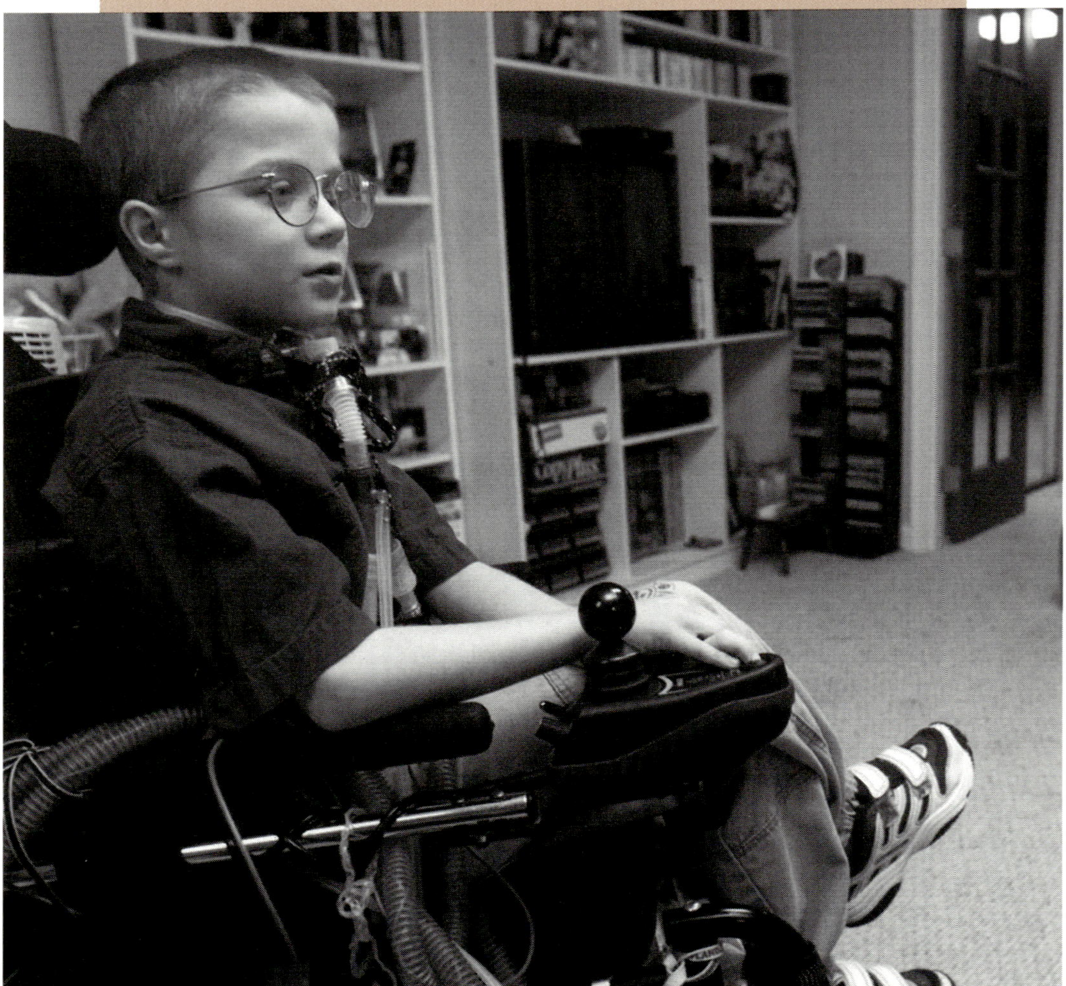

Mattie Stepanek, shown here at age eleven, has fought a battle with muscular dystrophy most of his life. In spite of his disease, he is an accomplished poet.

Types of Muscular Dystrophy

There are many different types of muscular dystrophy, including the following.

- The most common type of MD to affect adults is *myotonic* (mye-oh-TAHN-ik) *dystrophy* (DM). DM has a symptom unique from other MDs: stiffening or spasms in the muscles after use. DM progresses slowly. It often first affects the muscles of the face, feet, hands, and neck.
- *Duchenne* (doo-SHEN) *muscular dystrophy* (DMD) is the most common—and the most severe—type of MD. DMD usually first appears in children between two and five years old. The condition progresses very rapidly, and by the age of twelve, most children with DMD have to use wheelchairs. DMD usually results in a shortened life span.
- *Becker muscular dystrophy,* which often appears in adolescence, progresses more slowly than Duchenne. Although patients may eventually lose the ability to walk, they have a much greater chance of living a longer life.
- *Limb-girdle muscular dystrophy* has been known to first appear in people as young as five and as old as thirty. There are at least ten different types of limb-girdle MD. This type of MD normally progresses slowly. People usually lose most of their ability to walk about twenty years after its onset.
- *Facioscapulorhumeral* (fash-ee-oh-skap-yoo-lor-HYOO-mer-ul) *muscular dystrophy* (FSH) usually appears during adolescence. It first affects the muscles of the face and then progresses to the shoulders, back, pelvis, and legs. Although the disease usually progresses slowly, there may be periods when it suddenly becomes much worse. About half of all people with FSH will lose the ability to walk at some point in their lives.

AUTOIMMUNE DISORDERS

An autoimmune (aw-toh-im-MYOON) disorder occurs when the body's immune system attacks other parts of its own body that it mistakenly identifies as foreign or harmful. Sometimes, these attacks and the resulting conditions are triggered by agents in a person's environment, such as a chemical to which the person is exposed or an illness or fever. In addition, a person who develops an autoimmune disorder may already have a special genetic trait that predisposes him or her to develop the condition.

MYASTHENIA GRAVIS

One autoimmune disorder that affects the muscle system is *myasthenia gravis* (mye-uss-THEEN-ee-uh GRAY-viss). Myasthenia gravis causes the skeletal muscles to weaken. This weakness becomes worse with activity and improves with rest. The condition results when messages from the nerve cells that control voluntary muscles do not reach the muscles properly. Myasthenia gravis can affect people of all ages and in all areas of the world.

Myasthenia gravis often affects the facial and neck muscles. People suffering from this condition may experience double vision, as well as drooping eyelids, as their facial muscles weaken. Another symptom of the disorder is difficulty chewing and swallowing due to weakened throat muscles. Myasthenia gravis may eventually spread to arm and leg muscles, causing difficulty with daily activities. Myasthenia gravis can become life threatening when it interferes with a person's ability to breathe.

Although there is no known prevention or cure for myasthenia gravis, patients can control the condition by carefully pacing their physical activity, resting frequently, and eating a healthy diet. Foods that are high in potassium (bananas, tomatoes, oranges, and broccoli) are particularly good for people with this condition.

Fast Fact

The term *myasthenia gravis* comes from Latin and Greek words meaning "serious muscle weakness."

MYOSITIS

Another type of autoimmune disorder that affects the skeletal muscles is *myositis* (mye-oh-SYE-tiss). Also called inflammatory myopathy, this disorder causes swelling and eventual degeneration of the body's voluntary muscles. There are a number of different types of myositis. Some are more serious than others.

DIAGNOSING MUSCLE DISORDERS

Doctors use a number of methods to test for muscle injuries and diseases. An *electromyograph* (ee-lek-troh-MYE-uh-graf), or EMG, is a test used to diagnose muscle problems. An EMG tests how well muscles react to electrical stimuli. Doctors insert a needle electrode through the skin and into the muscle. The patient is then asked to move the muscle being tested. The muscle's response is displayed on a special machine called an oscilloscope (ah-SILL-uh-skope).

Blood tests can be used to detect antibodies that are present in people who have certain autoimmune conditions that affect the muscles. *Antibodies* are proteins that are produced by the body to attack bacteria and other foreign invaders. Myasthenia gravis, for example, may be diagnosed with a blood test. Doctors can also perform muscle biopsies, taking tissue directly from a muscle to be tested. Lastly, genetic tests, using blood samples, can be performed if a condition is thought to be genetically inherited.

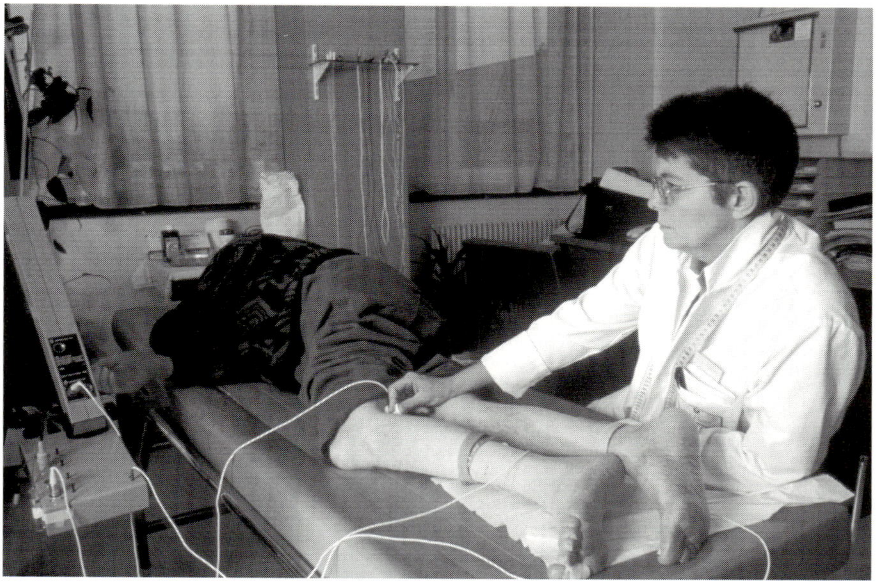

A patient undergoes an electromyograph (EMG) while a physician monitors his muscle responses on an oscilloscope.

New York Yankees first baseman Lou Gehrig, nicknamed the "Iron Horse," wipes away a tear at the famous 1939 ceremony celebrating his baseball achievements. Gehrig's career was cut short by amyotrophic lateral sclerosis (ALS).

OTHER MUSCLE-RELATED DISORDERS

Doctors are still working to discover the causes of some muscle conditions. One such disorder that affects the muscles is *amyotrophic lateral sclerosis* (uh-mye-uh-TRAH-fik LAT-er-ul skluh-ROH-siss), or ALS. This disease is also known as "Lou Gehrig's disease," after baseball player Lou Gehrig (1903–1941). Doctors do not know what causes ALS. This condition affects the neurological system, attacking nerve cells that control skeletal muscle function. ALS is a progressive and fatal disease. This means that ALS gets worse with time and eventually causes death.

Polio (POH-lee-oh) is another disorder that affects the muscles. This *infectious,* or contagious, disease is caused by a *virus,* a simple germ organism. The virus enters the bloodstream and may affect the central nervous system, attacking the cells that control muscle movement. This can result in permanent paralysis.

Multiple sclerosis (skluh-ROH-siss), or MS, is an autoimmune disease that can cause wasting and paralysis of the muscles. This condition causes nerve tissues to become inflamed. Eventually, nerve impulses to various parts of the body, including the muscles, may become blocked.

4

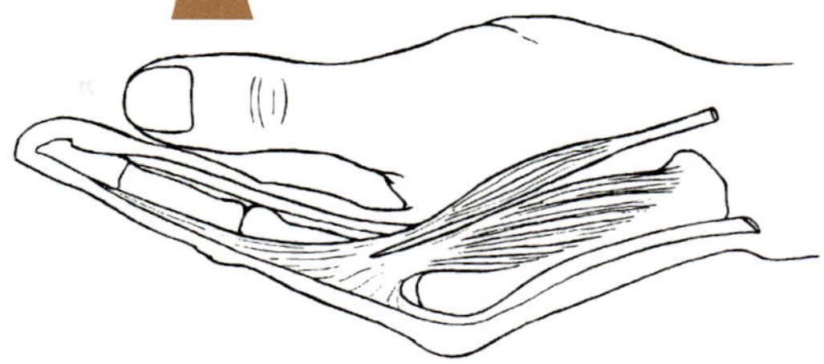

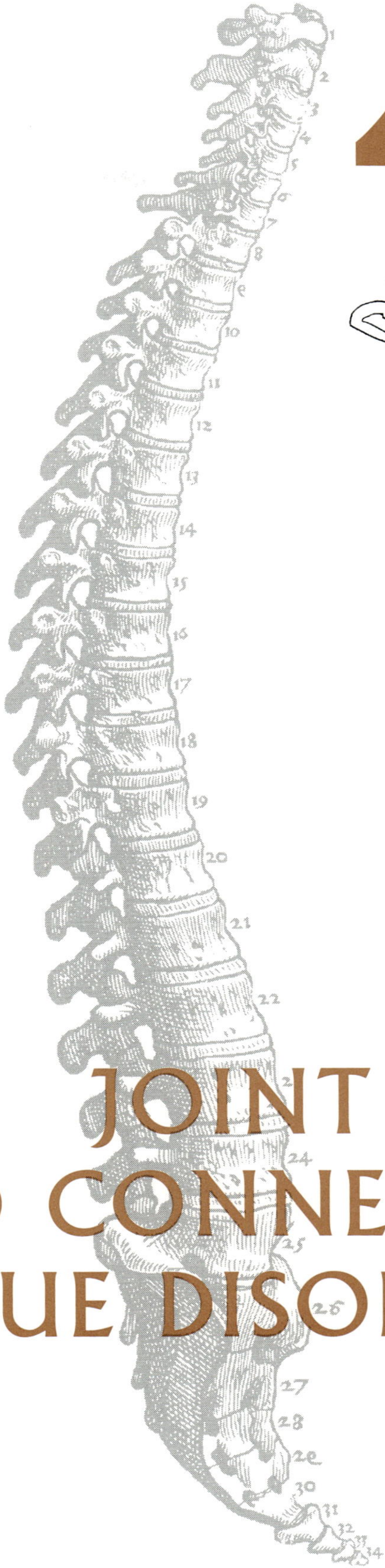

JOINT AND CONNECTIVE TISSUE DISORDERS

Conditions that affect the joints and connective tissues can cause pain, stiffness, and disability. Millions of Americans are affected by these types of problems.

ARTHRITIS

Arthritis (arth-RYE-tiss) is a painful stiffness and swelling in the joints. It can also cause pain to muscles, ligaments, tendons, and bones. There are more than 100 different kinds of arthritis. This condition can affect many different areas of the body.

Although doctors know what causes some types of arthritis, they are still in the dark about the exact causes of many other kinds. Some are caused by stress or repeated injuries. Others may be caused by genetic factors and environmental triggers. Still other types are autoimmune disorders.

Although many people think of arthritis as a disease that only affects the elderly. This is untrue. Arthritis affects people of all ages, from infants to senior citizens. In fact, about half of all those suffering from the condition are under the age of sixty-five.

TYPES OF ARTHRITIS

The most common type of arthritis is *osteoarthritis* (ah-stee-oh-arth-RYE-tiss). Osteoarthritis, also known as degenerative arthritis or degenerative joint disease, affects more than 20 million people in the United States alone. With this condition, cartilage within a joint or joints begins to wear down, usually as a result of aging, overuse, or repeated injury to the joint.

As the bones in the joint rub against one another, they can cause stiffness, pain, and a loss of mobility. In addition, *bone spurs,* also known as osteophytes (AH-stee-oh-fites), may grow on the joints. Sometimes, these small growths of bone break, causing more pain and swelling in the joint.

As people age, it is normal for their cartilage to wear down. As a result, most people over the age of sixty are affected by some degree of osteoarthritis. Some cases may be sped up by a person's genetic predisposition, injury, disease, or obesity. Osteoarthritis usually affects the weight-bearing joints in the body, especially the neck, lower back, hips, and knees.

A doctor examines a seventy-nine-year-old patient who has suffered from rheumatoid arthritis for over forty years.

The second most common type of arthritis is *rheumatoid* (ROO-muh-toyd) *arthritis.* Unlike osteoarthritis, rheumatoid arthritis is not caused by "wear and tear" on joints. Instead, rheumatoid arthritis is an autoimmune disease. Like osteoarthritis, this condition can be painful and debilitating.

Rheumatoid arthritis is an inflammation of the lining of a joint. It can cause pain, swelling, and even deformity of the affected joint. This type of arthritis usually affects the hands and feet, although it can also affect the entire body. Rheumatoid arthritis can spread to cartilage, ligaments, and tendons, as well, causing severe pain and stiffness. Chronic rheumatoid arthritis can affect the eyes, skin, heart, nerves, and lungs.

More than 2 million Americans suffer from rheumatoid arthritis. The majority of people affected are women. One particular type of the disease, known as juvenile rheumatoid arthritis, affects children. *Juvenile rheumatoid arthritis* is the most common type of arthritis affecting children.

At early stages, rheumatoid arthritis may be treated with medicines. Medicines can ease swelling and pain and may force the condition into *remission,* or temporary disappearance. Doctors recommend that people with rheumatoid arthritis stay physically active. An exercise program, developed with the assistance of a doctor, can help patients lessen the pain and keep their joints in good working condition. Walking and lifting weights are just two types of exercise that doctors say can help those suffering from rheumatoid arthritis. In severe cases, surgery may be necessary. Surgery can prevent the deterioration of the affected joints. Sometimes, the joint must be completely replaced.

Fibromyalgia (fye-broh-mye-AL-jee-uh) occurs in tissues that support bones and muscles. This type of arthritis most frequently affects the neck, spine, shoulders, and hips. In addition, people who have this disorder may suffer from sleep disturbances, irritable bowel syndrome, and anxiety. Doctors are not sure of the cause of this disorder—it might be injury, a virus, or even sleep disturbances.

Learning More about Lupus

Little is known about what causes lupus. Statistics show that women are more likely to be affected by the disorder than men are. Some racial and ethnic groups are more likely to develop the condition than others are. Researchers want to know why. NIAMS has funded a major lupus study to try to solve some of the mysteries surrounding the disease. The group has also started lupus registries by collecting tissue and blood samples from people with lupus and from their families. By studying these samples, researchers hope to learn more about the links among people who have the disease.

So what have researchers discovered? There may indeed be a genetic link to lupus. However, the specific gene that might cause the condition has not been found yet. Researchers are also still looking for the environmental trigger that might cause the lupus gene to become active. Scientists and medical professionals hope that with this new information, they can create a drug to cure lupus. Hopefully, the future will bring greater understanding about this painful condition.

Systemic lupus erythematosus (LOOP-us eh-rith-uh-muh-TOH-siss), or lupus, is an autoimmune disorder that causes arthritis. In addition to affecting the body's joints and muscles, lupus may also affect the skin, kidneys, heart, lungs, blood vessels, and brain. Symptoms of this condition include pain and swelling of the joints, skin rashes, chest pain, fever, fatigue, nausea, and seizures.

Doctors are not exactly sure what causes lupus. There is presently no cure for the condition. People with lupus go through periods of illness and wellness. Some lupus symptoms, such as arthritis, can be treated with drugs.

In general, people with lupus can live normal, healthy lives when they work with a doctor to learn all that they can about the condition and find a proper course of treatment. With knowledge and attention to their symptoms, patients can learn to ward off flare-ups or at least decrease the strength of the flare-ups. In the most serious cases, when lupus affects the heart, lungs, or nervous system, it can cause death.

Another type of arthritis is *ankylosing spondylitis* (AN-kil-oh-sing spon-dih-LYE-tiss). This form of arthritis is an autoimmune disorder that causes inflammation of the tendons and ligaments. It most often affects the joints between the vertebrae in the spine. The condition can also cause pain, swelling, and tenderness in the hips, shoulders, and knees. Ankylosing spondylitis usually affects older adolescents or young adults. Men are twice as likely to get the condition as women.

Is It Arthritis?

People with one or more of the following symptoms should check with their doctors to see if they might have arthritis.

- swelling in a joint or joints
- stiffness of the joints in the early morning that lasts for longer than one hour
- constant, recurring pain, tenderness, or warmth in a joint
- difficulty moving or using a joint
- redness of the skin around the affected area

TREATING ARTHRITIS

Doctors say that those suffering from most types of arthritis can ease their pain by staying active and healthy. According to the Arthritis Foundation, exercise is an important part of treating the disorder. Exercise strengthens joints and improves their flexibility. At the same time, it also strengthens muscles, allowing them to better protect the joints. The Arthritis Foundation points to gardening, hiking, yoga, and water exercise as just a few of the many activities that can help people stay in shape and feel better.

In addition to exercise, NIAMS recommends that people with arthritis eat a proper diet. It is especially important for them to keep their weight down. Excess weight places extra strain on the body's weight-bearing joints.

A number of drugs ease the pain of arthritis. Aspirin is one over-the-counter drug that has been used for many years. An *anti-inflammatory drug,* aspirin eases the swelling in joints. It also lessens the pain caused by arthritis and other joint conditions. Prescription drugs may also be used to reduce swelling and ease pain.

Members of the Grand Island YMCA Arthritis Aquatic Program exercise regularly in the pool to strengthen muscles and relieve some of the painful symptoms of arthritis.

SPRAINS, STRAINS, ACHES, AND PAINS

Sprains, strains, aches, and pains are common. Injuries that affect the joints and connective tissues can happen to anyone. However, some people are more likely than others to experience these types of injuries. Athletes and the physically active, as well as people in certain professions, are more prone to certain types of joint and connective-tissue injuries.

SPRAINS AND STRAINS

Sprains are usually stretched or torn ligaments. Although ligaments are elastic and flexible, they can only stretch so far. People suffer sprains as a result of a fall, twist, or other injury. According to NIAMS, ankle sprains are the most common injuries in the United States today. Ankle ligaments are often sprained during running, jumping, or sliding activities. Other common sprains affect the knees and the wrists. Sprained wrists often occur during a fall, when people put out their hands to try to protect themselves.

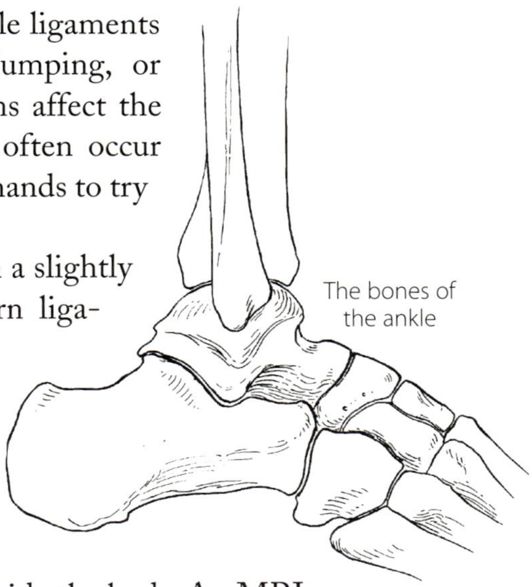

The bones of the ankle

A sprain can range in severity from a slightly stretched ligament to a completely torn ligament. Symptoms of a sprain include pain, swelling, numbness, bruising, and the inability to move the joint. To diagnose a sprain, doctors may use an X-ray or *magnetic resonance imaging* (MRI) to examine the injury. These two machines take pictures of what is inside the body. An MRI is more detailed than an X-ray.

Strains occur when tendons or muscles are damaged, usually by twisting or pulling motions. Strains can occur when people don't warm up properly before exercising. Strains often occur in the hamstring muscles, in the back of the thigh. Contact sports, including football, soccer, hockey, boxing, and wrestling, increase the risk of this type of injury, but those who participate in other sports are also at risk. Symptoms of strains include pain, muscle spasms, and muscle weakness. Cramping and swelling may also cause problems.

Another common type of strain is to the lower back, or *lumbago* region. Aches, pains, and stiffness in this area are known as lumbago (lum-BAY-goh). Lumbago can be caused by strains in the tendons and/or the muscles of the back.

The Achilles (uh-KILL-eez) tendon is also affected by strains. The Achilles tendon runs along the back of the calf. Jumping places a great amount of stress on the Achilles tendon. As a result, strains are common in those who participate in track, tennis, and basketball. A person who suffers a torn Achilles tendon will not be able to lift the heel. Although the injury often heals with rest and physical therapy, surgery may be necessary for more severe tears.

TREATING SPRAINS AND STRAINS

Immediately after a strain or sprain injury, patients are first treated with RICE. RICE is an acronym that doctors use for rest, ice, compression, and elevation.

RICE

- **REST:** Rest means taking it easy. Patients must cut down on exercise, sports, and other strenuous daily activities. It may even be necessary for a patient to keep all weight off the injured area by using crutches or a cane.
- **ICE:** Applying an ice pack to the affected area will reduce pain and swelling. Ice can be applied for twenty minutes at a time, from four to eight times a day. Any more than this might result in frostbite.
- **COMPRESSION:** Compression means wrapping or putting pressure on the injured area. Compression may include a cast, splint, or wrap to reduce swelling.
- **ELEVATION:** The injured person should keep the affected area *elevated,* or raised, above heart level to reduce swelling. Those suffering from sprains or strains can use pillows to get their affected area to the right height.

A person with a sprain or a strain may also want to take aspirin or another anti-inflammatory medication to ease the pain. For more serious strains or sprains, patients may be required to wear a cast until the affected ligament, tendon, or muscle has healed. Severe sprains and strains may even require surgery to repair any damage.

Eventually, a person who has suffered a sprain or strain will need to exercise the affected area to get it back into top shape.

OTHER JOINT DISORDERS

Some joints in the body are prone to problems. One such joint is the *temporomandibular* (tem-pur-roh-man-DIB-yoo-lar) *joint.* This joint is located where the lower jawbone, or *mandible,* connects to the temporal bone of the skull. The temporomandibular joint is one of the most frequently used joints in the body. We use it when talking, chewing, and yawning.

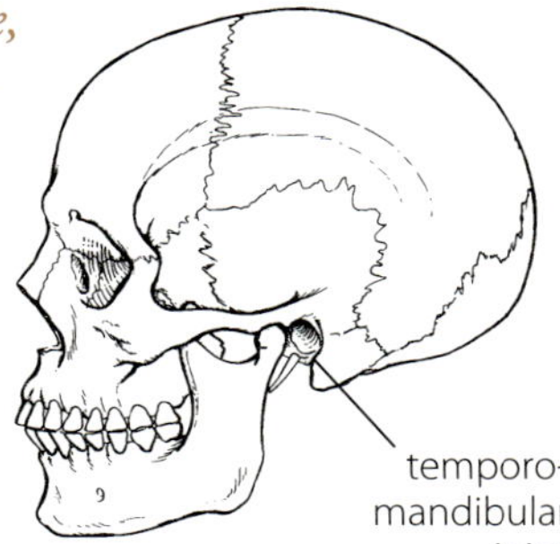

Temporomandibular joint (TMJ) disorders can result in aching jaws, ear pain, headaches, ringing in the ears, and popping and clicking sounds. In addition, the jaw may, at times, become locked. These conditions can be caused by injury, aging, normal wear and tear, and even excessive gum chewing! Other cases are caused when people frequently clench their jaws or grind their teeth. TMJ disorders may also be caused by what dentists call a malocclusion (mal-uh-KLOO-zhin). *Malocclusion,* or a "bad bite," occurs when the top and bottom sets of teeth do not fit together properly.

Most TMJ disorders are temporary. To ease the pain associated with the condition, doctors advise patients to rest, use ice and heat treatments as necessary, avoid chewing gum and hard foods, and avoid clenching or grinding their teeth. Pain medication can also ease aching and discomfort.

People who unconsciously grind or clench their teeth may use a splint or bite plate. This appliance, similar to a mouth guard, can be worn while sleeping, driving, or at other times. It takes the pressure off the teeth and jaw joints. Patients who have TMJ as a result of a bad bite may need dental work to solve the problem.

Other problems can result from injured joints. One such problem is *avascular necrosis* (ay-VASS-kyoo-lur nek-ROH-siss), the temporary loss of blood flow into bone. Long bones are most often affected. The loss of blood flow may be the result of an injury, the use of steroids, the excessive use of alcohol, or other factors. Without blood, the bone tissue dies and begins to collapse. Avascular necrosis can result in pain to the affected joint, followed by increasing pain if the joint and bone collapse. Drugs, exercise, and even surgery may be necessary to treat this problem.

SPINAL AND NECK CONDITIONS

Serious injuries to the spinal cord can result in paralysis. Each year, 10,000 people in the United States suffer spinal cord injuries. According to the Insurance Institute for Highway Safety, more than half of all traumatic spinal cord injuries are the result of motor vehicle accidents. Other causes include sports injuries, acts of violence, and falls.

In recent years, lawmakers have tried to reduce spinal cord and other injuries caused by motor vehicle accidents by passing laws that require people to wear seat belts in cars. Lawmakers have also asked car companies to design safer cars.

Fast Fact

According to the American Association of Neurological Surgeons, 80 percent of all spinal cord injury victims are male. Most are in their teens or late twenties.

A common type of spinal injury that does not result in paralysis is *whiplash.* Whiplash occurs when the neck is suddenly snapped forward and then backward. This snapping action may strain the ligaments or dislocate a cervical joint. Whiplash is treated with ice, followed by pain-relief medications, gentle therapy and exercise, heat treatment, and massage.

Sciatica (sye-AT-ih-kuh) is a pain that starts in the buttocks and radiates down the back of the thigh. This condition is caused by slipped cartilage disks in the spine pressing on the sciatic nerve in the spine. An operation may be necessary to relieve the pressure from the nerve.

Occupational Disorders

According to the National Institute for Environmental Health, 137 people die each day from job-related illnesses or injuries. Many more are injured while on the job. According to environmental health expert Dade W. Moeller (1927–), 25 percent of all workplace injuries are caused by lifting or moving objects. Another 15 to 20 percent are caused by slips and falls.

REPETITIVE STRESS INJURIES

Repetitive stress injuries (RSIs) are injuries that happen when parts of the body are stressed over and over. For this reason, they are sometimes called overuse injuries. These types of injuries usually affect the musculoskeletal system. Some RSIs occur as the result of job-related actions. Then they are known as occupational disorders. In children, RSIs are often the result of sports-related incidents. Other activities that may cause RSIs include using computers and playing video games. Symptoms of repetitive stress injuries include tingling, numbness, or pain; swelling, inflammation, burning, stiffness, or soreness, especially in the neck or back; persistent weakness or fatigue in the hands or arms; and frequent headaches. People who suspect that they are suffering from RSIs should seek immediate treatment from a medical professional.

What Causes RSIs?

- **REPETITIVE MOTIONS:** Repeating the same motions over and over can stress muscles and tendons and weaken them. The severity and location of the injury depends upon the type of movement, the muscles and tendons involved, the speed of the movement, and the force of the movement.
- **FORCEFUL MOTIONS:** The amount of physical effort required to do a job can contribute to a repetitive stress injury.
- **AWKWARD OR UNCHANGING POSTURE:** Repeated or prolonged reaching, twisting, bending, kneeling, or squatting can cause problems. Certain positions can also be harmful, such as working with the hands and arms above the head for long periods.
- **CONTACT STRESS:** Pressing the body against hard or sharp edges or surfaces puts pressure on tendons and causes problems.
- **VIBRATIONS:** Repeated exposure to vibrating movements can damage tendons and other parts of the body.

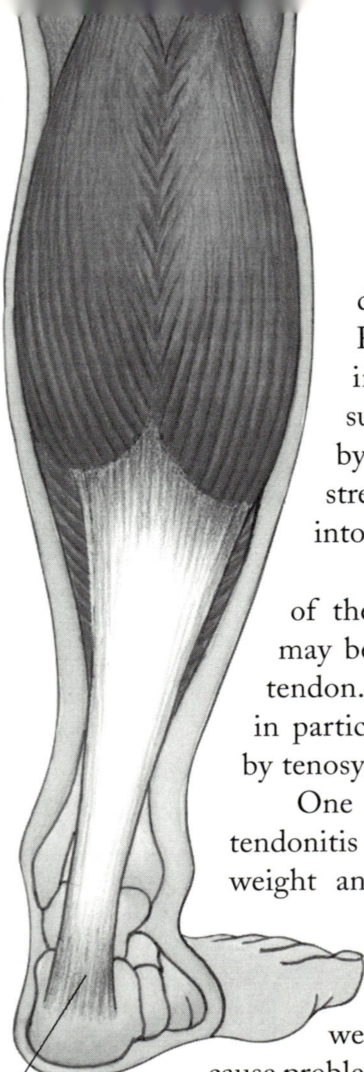

The Achilles tendon is the largest tendon in the body.

TENOSYNOVITIS AND TENDONITIS

Tenosynovitis and tendonitis are painful, debilitating conditions. Both can sometimes be RSIs. *Tenosynovitis* (ten-oh-sye-noh-VYE-tiss) is an inflammation of the synovial sheath that surrounds a tendon. Tenosynovitis is often caused by repetitive motion. It can also be caused by over-stretching. Tenosynovitis may eventually develop into tendonitis.

Tendonitis (ten-duh-NYE-tiss) is an inflammation of the tendon itself. Like tenosynovitis, tendonitis may be caused by repetitive motions or overuse of a tendon. Certain athletes are prone to the condition in particular tendons. Tendonitis may be accompanied by tenosynovitis.

One area of the body that is especially at risk of tendonitis is the foot. The foot bears large amounts of weight and pressure, and tendon damage here is not unusual. Running, jogging, kicking, and other activities can cause tendons to become inflamed. Even something as simple as wearing uncomfortable or ill-fitting shoes may cause problems. Another difficult area is the shoulder. Tennis and other racket sports can cause tendon damage here.

TYPES OF TENOSYNOVITIS AND TENDONITIS

There are a number of different types of tenosynovitis and tendonitis. *De Quervain's tenosynovitis* affects the tendons of the thumb. This condition, which is often the result of repetitive grasping or pinching motions, causes swelling, pain, and tenderness at the base of the thumb. This type of tenosynovitis is sometimes called trigger thumb.

Rotator cuff tendonitis is an inflammation of the muscles of the upper arm. People with this condition often feel shoulder pain that may continue down the entire arm. Other symptoms include stiffness and weakness that limit movement. This type of tendonitis can be caused by repetitive actions in which the elbows are held above head level. It may also be caused by excessive elevation of the arm and by aging.

Epicondylitis (ep-ee-kahn-dih-LYE-tiss) is an inflammation of the joints of the elbow. *Lateral epicondylitis,* also known as tennis elbow, is an inflammation of the outer tendons of the elbow joint. The condition earned its nickname because nearly half of all tennis players suffer from it at one time or another. However, only about 5 percent of all those who have lateral epicondylitis are actually tennis players. Others at risk for lateral epicondylitis include carpenters and gardeners. Those suffering from this RSI will feel pain on the outside of the elbow.

Medial epicondylitis is an inflammation of the inner tendons of the elbow joint. Those suffering from this type of epicondylitis feel pain on the inside of the elbow. The condition is nicknamed "golfer's elbow" because many golfers are affected by it.

Stenosing (sten-OH-sing) *tenosynovitis,* sometimes known as trigger finger, occurs when the surface of the fingers' tendons becomes irritated, rough, and thickened. People with this condition find it difficult to bend their fingers after they have been held straight for a while. When the finger or thumb is flexed, an audible popping sensation may occur. Stenosing tenosynovitis can be caused by the repetitive usage of tools that have sharp or hard edges.

BURSITIS

Bursitis (bur-SYE-tiss) is an inflammation of the bursae, the fluid-filled sacs in the joints. The bursae act as cushions to protect bones and joints. They can become irritated or damaged from repeated pressure on a joint. Bursitis commonly affects the hips, knees, elbows, and wrists. It causes swelling and pain. Luckily, bursitis can be easily treated by taking anti-inflammatory drugs and resting the affected joint.

Fast Fact

Health professionals say that people who frequently carry overloaded backpacks are at risk of bursitis and other musculoskeletal injuries. Many doctors recommend that kids carry no more than 10 to 15 percent of their body weight in their backpacks.

Bursitis may be caused by overhead lifting and the overuse of certain joints. People in some occupations are more prone to bursitis than others. Carpet layers, for example, are at risk for bursitis of the knee joints. The wear on the joint from kneeling on the hard surface of a floor irritates the tendon of the knee. As a result, bursitis of the knee is sometimes known as carpet layer's knee or housemaid's knee. Doctors and health experts recommend that people who spend a lot of time kneeling should wear protective kneepads to cushion the pressure to the knee joints.

CARPAL TUNNEL SYNDROME

Carpal tunnel syndrome is a swelling of the tendons or ligaments that run through the carpal tunnel, the space between the wrist bones. This swelling puts pressure on the median nerve that also runs through the tunnel. As a result, people suffering from carpal tunnel syndrome experience pain and numbness in the wrist and hand.

Carpal tunnel syndrome is known as an RSI because it is caused by repeated hand movements that irritate the wrist area. People who perform tasks that require them to repeatedly move their wrists up and down or keep their wrists bent while pinching or gripping are at risk from the condition. Occupations with a high incidence of carpal tunnel syndrome include computer users, meat cutters, carpenters, and mechanics. Such conditions as pregnancy, diabetes, and thyroid (THYE-royd) disease may increase the chances for developing carpal tunnel. These conditions cause people to retain fluids and raise the risk of inflammation and swelling.

People with carpal tunnel syndrome usually experience numbness and pain at the base of the thumb and the first three fingers as swollen tendons press on the median nerve. They may also notice a burning, tingling, or numbness in their fingers. The condition may also cause a pain that shoots up the arm as high as the shoulder. Carpal tunnel symptoms often first appear at night.

Doctors treat carpal tunnel syndrome with splints or braces, rest, anti-inflammatory drugs, and hand and wrist exercises. In some cases, doctors inject a substance called *cortisone* into the carpal tunnel. Cortisone is a hormone that reduces inflammation of the synovial membranes in the wrist. It is used to treat arthritis and other diseases. (*Hormones* are chemical messengers produced by the endocrine system that carry information from one part of the body to another. Your body needs hormones to keep it functioning properly.) If these methods of treatment do not work, however, surgery to relieve pressure on the nerve may be necessary. With proper treatment, most people recover completely from carpal tunnel syndrome.

In recent years, carpal tunnel syndrome has become more widespread. Each year, one out of every 1,000 people is diagnosed with the condition. If it is ignored or not diagnosed correctly, carpal tunnel syndrome can lead to permanent nerve damage.

To reduce the number of sufferers, computer mice, keyboards, and other workplace tools and machines have been redesigned to reduce stress on the tendons of the wrist. In addition, many employers now try to educate their workers on how to avoid the disorder. Tips that help include taking frequent rests, doing hand exercises, and working with the proper posture.

Kids and Repetitive Stress Injuries

Repetitive stress injuries usually affect adults, but kids can develop such conditions, too. More and more kids are using computers and playing video games. Because most computers and computer equipment are designed with adults in mind, kids face a greater risk of RSIs.

Doctors have even nicknamed one condition that they are seeing more frequently "Nintendo thumb." This RSI is a swelling at the base of the thumb, caused by the excessive playing of video games.

Two young players battle it out in a head-to-head video game. Repetitive use of the thumbs to play hand-held games can cause repetitive stress injuries.

Here are some steps that kids can take to avoid aches and pains caused by computer and video use.

- Proper posture is important. Make sure that you sit up straight, with your legs out straight and your feet flat on the floor in front of you. If your feet don't touch the floor, prop something beneath them—try your chemistry book! Don't slouch or lean over the keyboard.
- Don't hit the keyboard hard with your fingers. It's not necessary and can cause problems.
- When you type, keep your fingers and wrists level with your lower arms.
- Ask your parents to purchase a trackball to replace your mouse. Trackballs can prevent damage by allowing you to use more than one finger at a time.
- If you begin to feel any aches or pains, take a rest. Your body is trying to tell you something. Kidshealth.org recommends that you take a break after thirty minutes of computer time.

OSGOOD-SCHLATTER DISEASE

Osgood-Schlatter disease (OSD) is an RSI that often affects teens whose bones are still growing. It occurs when the tendon that connects the patella (kneecap) to the tibia becomes inflamed. The condition causes pain and swelling of the knee that worsens with exercise and other physical activity. In addition, fragments of the tibia may break loose and cause more pain.

BACK INJURIES

A number of back injuries can result from repetitive pulling and straining. Once back muscles and ligaments are scarred or weakened from these activities, this increases the risk of other such injuries. Improper or repetitive lifting of heavy items often causes back injuries on the job.

"Trucker's back" is an RSI that is caused by sitting in an upright position for many hours without the proper support for the small of the back. This is a common injury among long-haul truckers who don't use comfortable seats.

DIAGNOSING MUSCULOSKELETAL PROBLEMS

There are a number of ways that doctors can discover whether or not a patient is affected by arthritis or other joint and connective tissue conditions. One way is to take an X-ray. An X-ray machine uses electromagnetic radiation to take a two-dimensional picture of what's inside the body. Doctors use this test to diagnose fractures, tumors, and degenerative bone conditions. However, exposure to radiation can damage some cells, and overexposure has been linked to cancer and other problems. For this reason, pregnant women should never be exposed to X-rays.

Another way to discover joint and tissue conditions is with *computerized tomography* (tuh-MAH-gruh-fee), or CT. A CT machine uses a thin beam of electromagnetic radiation to create a three-dimensional picture of what's inside the body. This provides a much more detailed image of the body than an X-ray does. Doctors use the CT machine to diagnose spinal conditions and other bone injuries. Because the CT uses radiation, people run the same risks as they do with X-rays.

An MRI uses magnets and radio waves to create an image of the inside of the body. An MRI is much more revealing than a CT scan or an X-ray. Doctors use it to diagnose torn ligaments, injured muscles, sports-related musculoskeletal injuries, and many other medical conditions. MRIs are very safe. Unfortunately, MRIs are more expensive than X-rays, so many people with little or no health insurance can't afford them.

Common Sports Injuries

Here are some of the musculoskeletal injuries that are common in children's organized sports.

- **FOOTBALL:** Sprains, strains, pulled muscles, torn ligaments, fractured bones, and back injuries; knees and ankles are most commonly injured
- **BASKETBALL:** Sprains, strains, fractures, dislocations, and dental injuries; ankles, knees, and shoulders are most commonly injured
- **BASEBALL:** Strains and fractures
- **GYMNASTICS:** Strains and sprains

With a *bone scan,* a radiotracer is injected into the patient's bloodstream. The tracer decays and emits radiation, which is photographed by a special camera. This method is used to diagnose a number of musculoskeletal conditions, including rickets, bone infections, and arthritis.

Other joint conditions are found through a biopsy or a joint aspiration. In a *joint aspiration,* a needle is used to draw synovial fluid from an affected or injured joint. The synovial fluid can then be tested and analyzed.

AUTOIMMUNE AND GENETIC DISORDERS

Autoimmune and genetic disorders may also affect the joints and connective tissues. These conditions include dermatomyositis and Marfan syndrome.

DERMATOMYOSITIS

Dermatomyositis (dur-muh-toh-mye-oh-SYE-tiss) affects the connective tissues, resulting in inflammation of the muscles and skin. This autoimmune disorder affects people of all ages, although it is most common in people between forty and sixty. Women are more frequently diagnosed with the condition than men.

Because dermatomyositis affects the muscles, people with the condition may have some difficulty when performing everyday activities, including sitting, standing, or lifting the arms over the head. Red or purple rashes commonly appear. Dermatomyositis may also affect the heart and lungs. Patients are often treated with drugs that *suppress,* or control, the action of the immune system.

MARFAN SYNDROME

Marfan syndrome is a genetic disorder that affects the connective tissues of the body. Marfan syndrome can also affect other body systems, causing problems with the heart, brain, spine, lungs, eyes, and skin.

Marfan syndrome can strike people of any age, race, or sex. It can range from mild to severe. When Marfan syndrome affects the skeletal system, it can cause a person to be unusually tall, slender, and loose-jointed. The person may have long arms, legs, toes, and fingers, as well as a long narrow face, an arched roof of the mouth, and crowded teeth. Marfan may also affect the spine and breastbone.

There is currently no test nor cure for Marfan syndrome. Doctors usually diagnose the condition after a thorough physical examination. For Marfan patients with skeletal problems, doctors may recommend braces or even surgery to correct severe deformations.

Several historical figures may have been born with Marfan syndrome. They include the ancient Egyptian pharaoh Akhenaton (c.1300 B.C.E.), Mary Queen of Scots (1542–1587), and U.S. president Abraham Lincoln (1809–1865).

This young man suffers from Marfan's syndrome.
The genetic condition is responsible for his tall, thin appearance.

Did Abraham Lincoln Have Marfan Syndrome?

The image of Abraham Lincoln—tall, slender, and long-limbed—is recognized by every American. Some historians and scientists say that Lincoln's distinctive frame was a result of Marfan syndrome. At 6 feet, 4 inches (1.9 meters), the lanky Lincoln towered over most of his contemporaries.

Those who believe that Lincoln had Marfan point to his long frame, stooped posture, shuffling gait, and vision problems. These are all symptoms of the condition. In 1960, researchers learned that a man diagnosed with Marfan shared an ancestor in common with Lincoln.

However, many researchers do not believe that the president had this disorder. They say that Lincoln was within the normal limits for a tall, thin man. They also point out that Lincoln was farsighted, while people with Marfan are nearsighted. Scholars and other experts continue to debate whether or not Lincoln suffered from the disease. One day, tests of the president's *deoxyribonucleic* (dee-ox-ee-rye-boh-noo-KLAY-ik) *acid,* or DNA, may put an end to this medical mystery. (DNA is a molecule that contains hereditary information for an organism.)

MISCELLANEOUS CONDITIONS

Doctors are still trying to uncover the causes of several conditions of the joints and connective tissues. One such condition is a *bunion,* an inflammation of the joint at the base of the big toe. When the joint here becomes inflamed, a large bulge occurs. The bursa in the joint fills with fluid, enlarging the bulge. The bones there may thicken, too. The bunion may eventually have to be removed surgically. Although bunions are sometimes caused by ill-fitting shoes, often the cause is never discovered.

Another condition with an unknown cause is a *wrist ganglion* (GANG-lee-on), a cyst that usually occurs on the back of the wrist. A *cyst* is a small, abnormal growth. The fluid-filled cyst appears on ligaments, tendons, or joint linings as a result of irritation to the affected area. These cysts usually go away with time, but painful ones may need to be lanced with a needle to remove fluid or surgically removed.

5

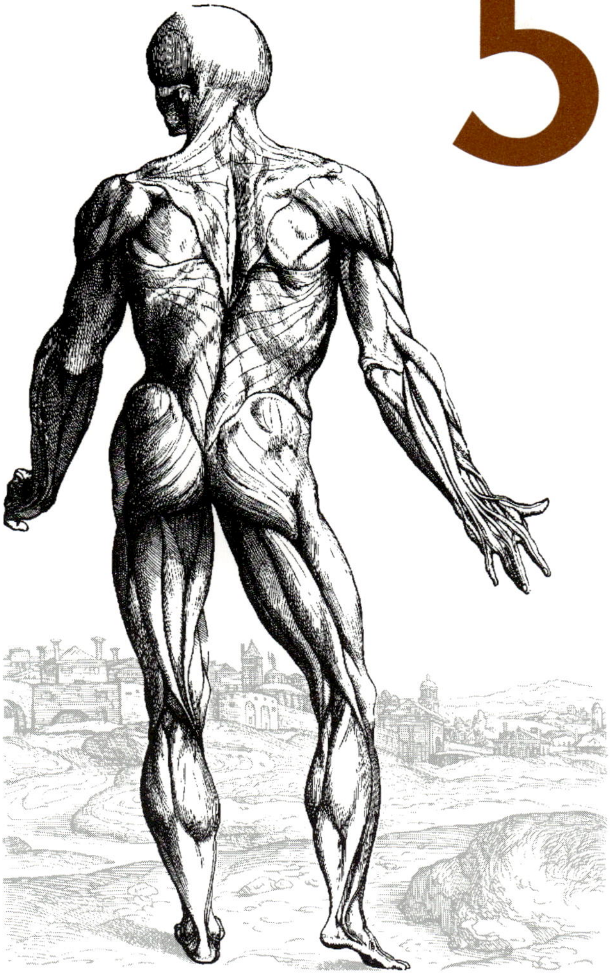

HOW THE ENVIRONMENT AFFECTS THE MUSCULOSKELETAL SYSTEM

There are many different ways that the environment can affect the musculoskeletal system. Chemicals, pollution, bacteria, viruses, and diet can all cause medical problems.

CHEMICALS AND DISEASE

There are hundreds of thousands of chemicals in the world. Some occur naturally in the environment, while some have been developed by humans. Some chemicals have been used to make the world a better place—to cure diseases, keep people healthy, and make our lives more comfortable. But chemicals can cause problems, too. Some chemicals can cause serious health conditions. Scientists don't even know the full effects that many chemicals may have on the human body.

Chemicals are all around us. We cannot avoid coming in contact with them every day. There are chemicals in the air that we breathe. Some chemicals are added to our drinking water, while others fall from the sky with rain and snow. These chemicals can end up in our soil and water supply. There are even chemicals added to our foods. It has been estimated that a person comes in contact with over 70,000 human-made chemicals during his or her lifetime, and the effects of most of these chemicals have not been studied. However, it is very important to try to understand the effects that chemicals can have on the human body.

Fast Fact

In 1945, Grand Rapids, Michigan, became the first city in the world to add fluoride to its water supply in an effort to reduce tooth decay. After eleven years, the rate of dental cavities among kids in the area had gone down by 60 percent.

SODIUM FLUORIDE

Some chemicals are good for you when taken in the proper doses but harmful if you take too much. In recent years, controversy has erupted over the benefits and potential risks of sodium fluoride. *Sodium fluoride* is a chemical that is commonly added to drinking water, foods, and toothpaste to prevent cavities and tooth decay.

However, sodium fluoride is also used in pesticides and other chemicals. Recent research has shown that too much sodium fluoride may trigger *fluorosis* (flor-OH-siss), a condition of the teeth that causes tooth enamel to become soft and porous. Fluorosis, which occurs as teeth are developing, may also cause tooth staining, pitting, and degeneration. Additionally, some researchers believe that too much sodium fluoride may lead to an increased risk of osteoporosis. Further studies are needed before a definite link can be established between the two, however.

Those most at risk for problems resulting from too much sodium fluoride are young children. They are more likely to eat or use large amounts of toothpaste. Nearly all health experts agree that sodium fluoride, taken in the proper amount, has a beneficial healthy effect. The daily recommended intakes of sodium fluoride for children up to the age of three years old is 0.1 to 1.5 milligrams (0.0000035 to 0.0000525 ounces). For children ages four to six, the daily recommended intake is 1 to 2.5 milligrams (0.000035 to 0.0000875 ounces).

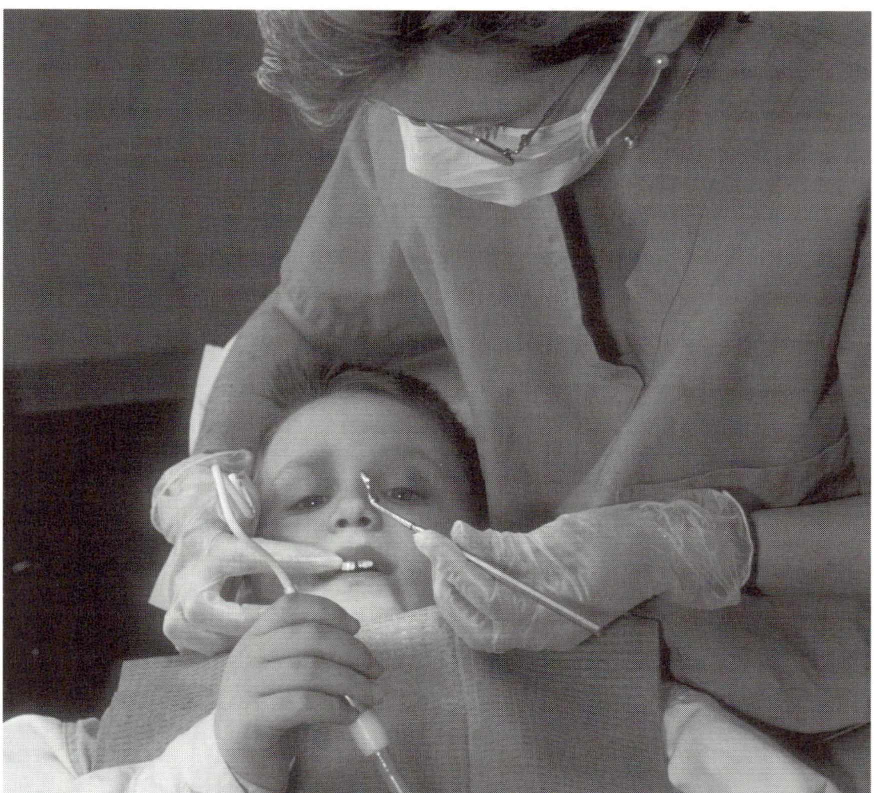

Regular treatments with sodium fluoride applied to the teeth of young children have been shown to reduce the instances of tooth decay.

Top Five Hazardous Substances

There are many hazardous substances in the world today. The Comprehensive Environmental Response, Compensation, and Liability Act (CERCLA) lists 275 chemicals that should be given top cleanup priority at hazardous waste sites around the nation. Here are the top five.

- **ARSENIC:** Arsenic occurs naturally in Earth's crust. Exposure to high levels can cause skin problems, including discoloration and small corns and warts. Arsenic can also cause throat and lung irritation when inhaled. If a lot of arsenic is ingested, death can result. Humans ingest arsenic by eating food, drinking water, or breathing air containing arsenic. Arsenic is used on wood as a preservative. Sawdust or smoke from such wood can also cause health conditions. Some areas have naturally high levels of arsenic in rock.
- **LEAD:** Lead also occurs naturally throughout the world. However, increased amounts of lead are put into the environment through human actions. Lead vapors are released into the atmosphere through the burning of fossil fuels and through mining and manufacturing. People who ingest lead-contaminated water, food, or vapors are at risk for some serious health problems. Lead poisoning can lead to anemia, muscle weakness, and even brain damage in children.
- **MERCURY:** Mercury is another naturally occurring metal. Most mercury exposure comes from breathing contaminated air or ingesting contaminated water and food. For humans, eating contaminated fish and shellfish is a source of methylmercury, a type of mercury that is commonly used to preserve grain. Mercury can cause serious health problems, from brain and kidney damage to birth defects. It can also cause lesser problems, including skin rash, eye irritation, nausea, vomiting, and an increase in heart rate and blood pressure.
- **VINYL CHLORIDE:** Vinyl chloride is a colorless, flammable gas that is used to make polyvinyl chloride (PVC). PVC is used to make many different plastic products, including pipes, wire and cable coatings, and automobile upholstery. Vinyl chloride can cause liver and nerve damage, as well as

liver cancer. Most people who are exposed to vinyl chloride breathe it in the workplace. The bones in the tips of the fingers break down in some workers.

- **POLYCHLORINATED BIPHENYLS (PCBS):** PCBs are mixtures of chemicals that were once used as coolants and lubricants in equipment. Although they have not been produced in the United States since 1977, PCBs still exist in the environment at some hazardous waste sites. They may also exist in products manufactured before 1977, including fluorescent lighting fixtures, electrical devices, and hydraulic oils. In children, PCBs can cause skin conditions and problems with the immune and nervous systems. PCBs have also been shown to cause cancer in laboratory animals.

Water wells in many villages in Bangladesh are contaminated with naturally occurring arsenic that leaches into underground aquifers. This contamination is a major health hazard for village residents.

LEAD IN BONE

Lead poisoning is one of the most serious environmental health hazards in the world today. Lead occurs naturally in the environment. It was also once used in paints, solder, gasoline, and glaze on cooking vessels. Lead poses an especially serious risk to small children. According to the CDC, an estimated 890,000 preschool-age children living in the United States had higher than acceptable levels of lead in their blood in the early 1990s. Children most often come in contact with lead by ingesting chips of lead-based paint that remain in some homes.

Lead can affect the human body in a number of ways. Lead in children's bones curtails proper growth. The CDC estimates that 4.4 percent of all U.S. children ages one to five have too much lead in their bodies. Lead can also cause hearing and learning problems.

Lead exposure is a problem for adults, too. In adults, lead is stored chiefly in bone. Lead can remain in bone marrow anywhere from seven to twenty years. For young women who want to have children, bone lead can be especially dangerous. During pregnancy, lead, along with calcium, is released into a woman's system and transferred to the developing fetus. The lead is then absorbed by the baby's developing bones and tissue. Lead can damage the baby's developing organs and body systems. Even after pregnancy, nursing mothers can transmit lead to their babies through breast milk.

Some researchers believe that there may also be a link between lead exposure and some cases of osteoporosis. Even exposure to lead as a child is believed to cause osteoporosis later in life. In 2002, scientists at the University of Rochester predicted that as a result of childhood lead exposure, the United States will witness an increase in lead-caused osteoporosis within the next decade.

A Burning Problem

Did you know that some candles emit lead when burned? When the wicks burn, the lead is released into the atmosphere. Lead residue is often left behind on walls, floors, and furniture.

In February 2001, the Consumer Product Safety Commission voted to ban all candles with lead wicks. However, some people may still have candles with these wicks in their homes. To check candles in your home, look at the tops of the wicks. If a wick contains lead, you will see it in the center of the wick.

LIVING WITH LEAD

The dangers of lead were first documented in 1904 by Australian scientist J. Lockhart Gibson (1794–1854). Even before then, health problems known as *lead choler* (KAHL-er) and *wrist drop* were common for painters, miners, and others who worked with lead. Despite the documented dangers of lead, the substance continued to be used. Beginning in the early 1970s, the United States took steps to ban the use of lead in paint and gasoline. In the late 1980s, lead solder was also banned. These measures have been successful. Since 1971, the levels of lead in blood have been reduced in the United States by 78 percent.

Although the use of lead in paint and solder has been banned, that doesn't mean that lead no longer exists in homes today. Millions of older homes still have pipes with lead solder and walls covered with lead-based paint. Lead from pipe solder can seep into drinking water. Also, as the paint on these walls ages and chips, small children may eat paint chips.

Many of the homes that contain lead paint and solder exist in low-income housing projects today. In fact, the Alliance to End Childhood Lead Poisoning says that children from low-income families are eight times as likely to have high levels of lead in the blood than kids from other families are.

Public *landfills* were also a problem in the past. Such landfills were dumping grounds where anything could be discarded. Lead, arsenic, and other metals and chemicals were dumped there in the past. From the landfills, the chemicals seeped into the water supply.

In recent years, landfills have been more closely regulated. Landfills must now have liners, groundwater monitoring, collection for chemicals that seep into the ground, and gas removal systems. In addition, landfills cannot be built near faults, wetlands, and floodplains.

LEAD SAFETY

How can people stay safe and avoid lead exposure and lead poisoning? Health experts recommend that houses built before 1980 should be tested for lead pipes and paint. Families who live in or want to buy such houses should look for chipping paint on walls and other painted areas. Even older playgrounds should be checked for flaking paint.

If leaded paint is found in the home, people may choose to paint over it with a nonleaded paint or a special sealant. However, for a long-term fix, the paint should be removed by a professional. Whenever possible, any household item covered with lead paint should be replaced.

To reduce the possibility of ingesting lead from old pipes, run the faucet for a minute or two each morning before using the water. In addition, use only cold water, not hot, for drinking and cooking.

Spotlight on Lead Contamination

Herculaneum, Missouri, is a quiet little town along the Mississippi River. It is considered a "disease cluster." Lead emissions from a nearby smelting factory, the largest such smelter in the nation, have polluted the town's air, soil, and water sources.

People who live in Herculaneum believe that lead pollution is making their children sick. Blood samples taken from children under the age of six in Herculaneum have shown that about 25 percent of them are suffering from lead poisoning. The national average for children under six with lead poisoning is about 3 percent.

To solve the problem, the Doe Run Company, which runs the smelter, has agreed to fund a multibillion-dollar cleanup effort. This includes replacing soil in contaminated yards and along contaminated roadways. It also includes cleaning the insides of people's homes if lead is found there. Contaminated churches, schools, and playgrounds must all be rid of the toxic substance, as well.

Some families with small children were relocated to homes in safer areas. Many families would like to move out of the town permanently. This is impossible for some, however, because they can't sell their homes in Herculaneum. No one wants to move to a town where their families cannot be safe.

The smelting company has also agreed to reduce emissions from its 110-year-old smelter. These efforts will be overseen by the Environmental Protection Agency (EPA), a government agency formed to protect human health and safeguard the natural environment. EPA officials warned area residents, however, that some lead emissions are unavoidable as long as the smelter is in operation. However, because the Doe Run Company employs 250 people from the town, many don't want it to close.

Opposite:
A sign on the street near the Doe Run lead smelter warns residents of Herculaneum, Missouri, of the dangers of high lead levels on surrounding streets.

AGENT ORANGE

Agent Orange was a herbicide used by the U.S. military in South Vietnam during the Vietnam War (1964–1975). A *herbicide* (HUR-bih-syde) is a chemical that is used to kill weeds and other plants. Agent Orange was used to destroy the forests where enemy soldiers could hide. One of the chemicals used to make Agent Orange was *dioxin* (dye-OKS-in), a substance known to cause severe skin problems and aggravate preexisting liver and kidney diseases.

After the war, Vietnam veterans began having numerous health problems. The incidence of these problems was much higher among veterans than among other U.S. populations. Health officials noted that veterans more frequently developed such diseases as Hodgkin's disease, non-Hodgkin's lymphoma, multiple myeloma, and chloracne.

Several of the conditions affecting Vietnam veterans affect the musculoskeletal system. Non-Hodgkin's lymphoma, for example, can form in the bone marrow. In addition, the children of veterans were at a higher risk of developing *spina bifida,* a birth defect that affects the spine. There may also be a link between Agent Orange and other birth defects.

United States Air Force planes spray the defoliant chemical Agent Orange over dense vegetation in South Vietnam in this 1966 photo.

In 2002, the United States and Vietnam signed an agreement to work together researching Agent Orange and its effects on the human body and the environment. Scientists from both countries will share research and results and will work together to decide what more needs to be done.

MERCURY

Mercury is a metallic element that occurs naturally in the world. It has several different forms. *Metallic mercury* is one form. At room temperature, metallic mercury is a silvery liquid. For this reason, it has been called quicksilver and liquid silver. Metallic mercury is used in thermometers, batteries, some paints, and some types of dental fillings. When this type of mercury is heated, it turns into a colorless, odorless gas. This gas is very dangerous to all living creatures. Metallic mercury has been found at more than 700 hazardous waste sites in the United States.

The other forms of mercury are organic and inorganic. *Inorganic mercury* results when mercury combines with elements such as sulfur, chlorine, or oxygen. A compound called *mercury salt* is formed. *Organic mercury* is formed when mercury combines with carbon. The most common type of organic mercury is *methylmercury* (meth-ul-MER-kyur-ee), which is produced by microscopic organisms in the water and soil. When released into the surrounding environment, methylmercury can build up in the tissues of fish. The older and bigger the fish, the more mercury is likely to be found in its body. Swordfish and shark contain more mercury than do other types of fish. Mercury is also released naturally into the environment through the erosion of rock, soil, and minerals, as well as through volcanic explosions.

> **Fast Fact**
>
> Scientists estimate that between one-third and two-thirds of all mercury released into the environment is the result of human actions.

Humans are commonly exposed to methylmercury when they eat fish and shellfish that have been contaminated with the substance. Eating contaminated seafood puts people at risk for serious illness. The human body cannot process mercury. In addition, mercury is slow to leave the body. Excessive levels of mercury in the body can cause damage to the brain and kidneys and can affect a developing fetus.

Fast Fact

Although the American Dental Association says that Americans have nothing to fear from amalgams, some dentists disagree. There are a number of dentists throughout the nation who are "mercury-free." They will not use amalgams to fill their patients' cavities.

MERCURY POLLUTION

Ingesting mercury through contaminated fish is the most common way people are exposed to it, but it is not the only way. Mining and the burning of waste and fossil fuels have greatly increased the levels of mercury in the world. When the element is heated, mercury vapors escape into the atmosphere. Many people are exposed to these vapors. Exposure to mercury vapors can cause skin rashes, vomiting, diarrhea, and lung damage.

Children who are exposed to high levels of mercury are particularly at risk for serious healthy problems. Some may develop *acrodynia,* also called pink disease. Children suffering from acrodynia may experience severe leg cramps, redness of the skin, rashes, and peeling of the skin on the hands, nose, and feet. In addition, mercury exposure can affect the normal growth and development of the brain. Children who were exposed to mercury while in the womb may be born with mental retardation, blindness, seizures, and other health conditions.

MERCURY IN THE NEWS

In August 2001, Maine became the first state to require dentists to warn people that they may be getting mercury in their next cavity filling. That's because many dentists use *amalgams* (uh-MAL-gumz), types of tooth fillings that contain mercury. These silver-colored cavity fillers are made up of different types of metals, with mercury making up about 50 percent of the filling.

As an amalgam erodes, very small amounts of mercury vapor may be released into the mouth. The CDC reports that people have little to fear from the tiny amounts of mercury vapor released by amalgam fillings. However, further studies on the subject are being conducted.

Mercury contamination in the Amazon region of South America has also become a hot topic. Gold miners there use mercury to separate gold from dirt. Liquid mercury is mixed in with soil that contains gold. The mercury and gold bond together and can be more easily separated from the dirt and water. Later, the gold-mercury mixture is heated. The mercury turns into a vapor, leaving behind the precious ore. During this process, miners breathe in the poisonous mercury vapors. The mercury used in the process has also contaminated local water and fish supplies.

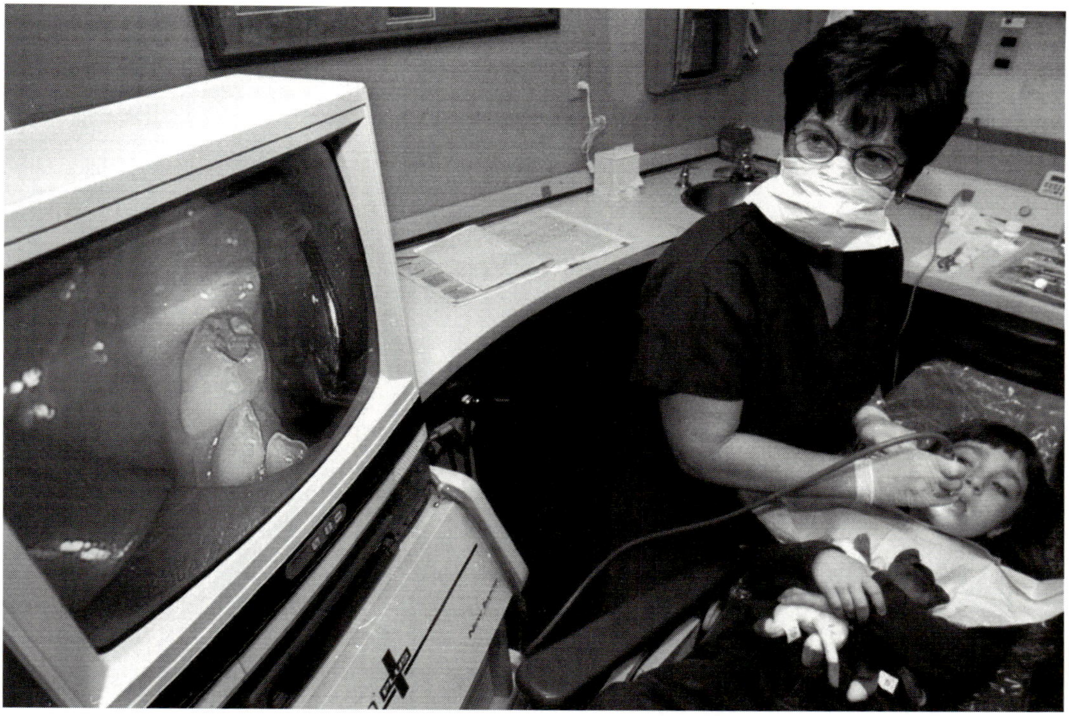

Using a small video camera, a dental hygienist looks at a television monitor to inspect a silver filling in this five-year-old's tooth. Some consumer groups contend that patients do not have enough information regarding the risks of using silver fillings containing mercury amalgam.

People who live near the mines have suffered the ill effects of the mercury used in the mines. In 1999, a number of people living in Amazon villages were affected by *Minamata* (min-uh-MAHT-uh) *disease,* also known as methylmercury poisoning. This disease attacks the immune system, causing shaking and severe muscle wasting. Minamata disease was first diagnosed in Minamata, Japan, in the 1950s. Today, many organizations have demanded an end to the use of mercury in gold mining.

Health and History: Cases of Mercury Poisoning

During the twentieth century, the use of mercury was widespread. Here are some examples of mercury poisoning in recent years.

- The first widespread case of mercury poisoning occurred in the 1950s in Minamata, Japan. Large numbers of people here were poisoned when they ate mercury-contaminated fish taken from a nearby bay. The bay had been polluted by a metal factory. Today, hundreds still suffer from the effects of Minamata disease. Some of those affected were children born to mothers who ate the contaminated fish from the bay waters.
- The practice of coating grain seeds with methylmercury to protect them from fungi had a deadly effect in Iraq during the early 1970s. In 1971 and 1972, grain treated with methylmercury poisoned thousands of people there. Although the coated seed was intended for planting, many people used the seed to make bread. The poisoned bread caused 6,500 Iraqis to be hospitalized with neurological problems, and more than 450 died.
- In 1969, a family of ten in Alamogordo, New Mexico, suffered mercury poisoning. They ate the meat of pigs that had been fed grain treated with methylmercury.

This woman is one of possibly tens of thousands of victims of mercury poisoning resulting from industrial dumping in Minamata Bay, Japan. Her legs are numb and do not function as a result of eating mercury-contaminated fish from the bay.

SAFETY TIPS

Humans are exposed to varying levels of mercury every day. However, it is important for people not to panic, but to take safety precautions to limit their exposure. For example, people should not stop eating commercial fish and seafood. The health benefits of seafood are well documented. In addition, the safety of commercial seafood in the United States is carefully monitored by the Food and Drug Administration (FDA), a federal agency that enforces U.S. government guidelines for foods and drugs.

Some groups do recommend that pregnant women limit their intake of shark and swordfish, two types of fish that contain more mercury than others. Canned tuna fish may also contain enough methylmercury to harm a developing fetus. Hunters and fishers taking wild game may want to check for local wildlife or fish advisories. These warnings let people know of high mercury levels in local wildlife.

Because of the potentially serious health effects of mercury buildup in the body, people should take care when handling items that contain mercury, including thermometers, barometers, batteries, fluorescent light bulbs, and some medicines. Small children should never be allowed to handle or play with these items. When they break, they should be carefully and safely disposed of.

In the home, large mercury spills should be reported to the health department. Do not try to clean up the mess yourself. It is especially important not to try to vacuum up spilled mercury. This will only cause it to turn into a vapor. People who believe that they have been exposed to high levels of mercury may want to be tested. There are currently blood, urine, and hair sample tests that can accurately measure levels of mercury in the body. The mercury will eventually leave the body over a period of weeks or months.

Multiple Chemical Sensitivity

Multiple chemical sensitivity (MCS) is a condition that may be triggered by exposure to chemicals. After the initial exposure, an allergic reaction can be set off by soaps, cosmetics, or almost anything. These reactions can cause muscle and joint aches, skin rashes, difficulty in breathing, memory loss, headaches, and nausea. MCS is a hotly debated condition. While some groups believe it is a serious health problem, others do not believe that MCS actually even exists.

DIET-RELATED CONDITIONS

The lack of a proper diet can cause serious health problems. While the incidence of diet-related disorders has decreased in some parts of the world, other areas continue to suffer. People who live in developing nations or those who live in poverty are more likely to suffer from diet-related conditions than other people are.

SCURVY

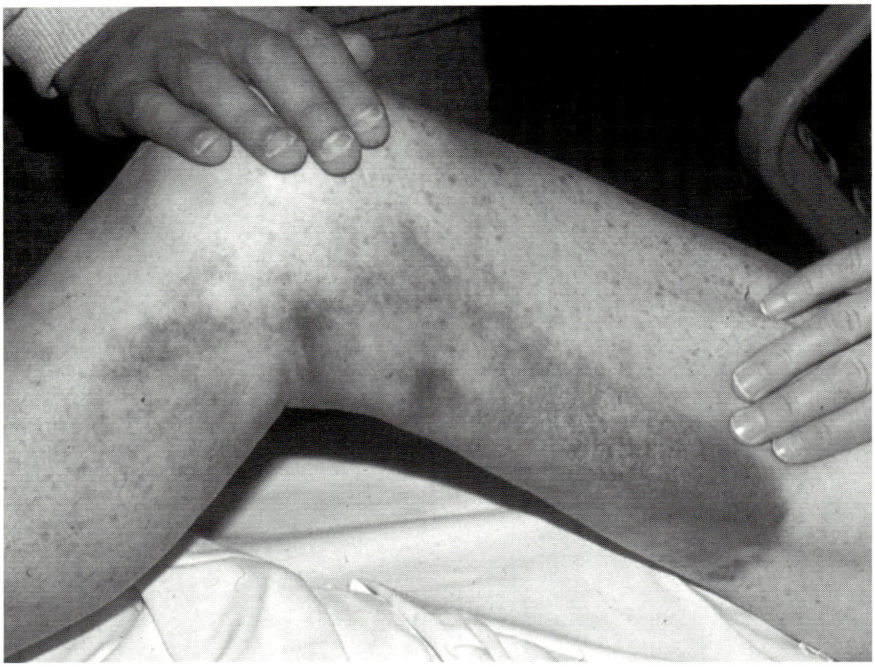

The effects of scurvy caused by a deficiency of vitamin C in the diet can be seen as swelling and inflammation of this person's leg.

Scurvy is a condition that results from a lack of vitamin C in the diet. Vitamin C contributes to the healthy growth of bones, organs, cartilage, ligaments, and blood vessels. Without this important nutrient, people may experience swollen and inflamed joints, anemia, gum disease, skin hemorrhages (bleeding), and loose teeth. Hundreds of years ago, scurvy was a serious problem for sailors on long voyages. Once a ship's supply of fresh fruits and vegetables ran out, seamen soon developed scurvy, earning it the nickname "the sailor's disease."

Today, scurvy is rare in the United States. Those who do get the condition are often older adults who don't have enough fruits and vegetables in their diets. It may also affect infants who are not fed nourishing foods or whose mothers did not eat a proper diet during late pregnancy. Other people at risk for scurvy include people who abuse alcohol and those who are unable to afford foods rich in vitamin C.

Scurvy sometimes causes problems in developing countries or those suffering from the effects of natural disasters. Disasters like droughts, hurricanes, and earthquakes deplete the supply of fresh food and water. In 2002, the WHO attributed forty deaths in Afghanistan to scurvy. At that time, the country was suffering from a drought.

Given daily doses of vitamin C, most people quickly recover from scurvy. Those who do not seek treatment, however, are at risk of death.

Scurvy and Sailors

For hundreds of years, scurvy plagued sailors around the world. Historians estimate that millions of seafarers lost their lives over the years. Ferdinand Magellan (c. 1480–1521) and Vasco da Gama (c. 1469–1524) both lost many crew members to scurvy during their historic voyages. On one of da Gama's voyages, more than half his men died of the disease.

Seamen had no idea that they were becoming ill due to a lack of fresh food. Centuries ago, before electricity and refrigeration, it was impossible to keep most meats, fruits, and vegetables from spoiling on long voyages. Although people realized that sailors were getting sick from their diet, they didn't know exactly what the cause was.

In 1747, Scottish naval surgeon James Lind (1716–1794) treated a group of twelve British sailors with different "cures" for scurvy. Some of the men received two oranges and a lemon daily. These sailors quickly recovered, while the others did not. Lind soon began giving the rest of the men the same treatment, and they, too, recovered. Lind believed that something in the oranges and lemons was helping the sick men. Before long, the British navy began supplying every ship that sailed out of port with fresh fruit and lemon and lime juice. This is how British sailors became known as "limeys."

Nearly 200 years after Lind's experiment, Hungarian-born scientist Albert Szent-Györgyi (1893–1986) discovered vitamin C in the spice paprika. In the 1970s, American scientist Linus Pauling (1901–1994) conducted research that highlighted the connection between vitamin C and a healthy body. Today, the benefits of vitamin C are well known.

RICKETS

Rickets is a condition that is caused by a lack of vitamin D in the diet. It mainly affects children. Children with rickets have weakness in and improper development of their bones, which may bow as they soften and bend out of shape. Curvature of the vertebrae and the legs show up as swelling at the ribs or knees. Another result of rickets is what is known as "pigeon breast." Here, the sternum becomes deformed and bows outward, protruding from the chest. Bones may also become brittle and break easily.

Rickets can be prevented and cured by making sure that people have enough vitamin D in their diets. Vitamin D is an important part of a healthy diet. It helps the human body use calcium and phosphorous, two nutrients that make bones, teeth, and nails strong and healthy.

In the past, many people took a daily spoonful of cod liver oil to avoid rickets. In the early 1930s, the United States began adding vitamin D to milk to eliminate rickets and other problems. People can also get vitamin D by eating foods that contain it. These foods include butter, eggs, and such oily fish as tuna and salmon.

The human body can get vitamin D another way: It can make it. The sun's ultraviolet (UV) rays trigger cholesterol (kuh-LESS-tur-awl) in skin cells to produce the vitamin. That's why vitamin D is sometimes called the sunshine vitamin. When the body has enough of the vitamin, the skin cells stop producing it.

During the Industrial Revolution in England, from the mid-eighteenth century to the mid-nineteenth century, smog from factories clouded the air, absorbing the UV rays needed to create vitamin D in the body. Children in factory areas became more prone to rickets. Even today, children in industrial areas are more prone to this disease.

Spina Bifida in the United States

In 1996, the FDA recommended that enriched grain products be fortified with extra folic acid. Two years later, the agency mandated that folic acid be added. Since then, the rate of babies born with spina bifida has greatly decreased.

SPINA BIFIDA

This young All-American Soap Box Derby racer is afflicted with spina bifida. Although he cannot use his legs, he is an able competitor in this race.

Spina bifida (SPYE-nuh BIH-fih-duh) is a birth defect in which the backbone and the spinal canal do not close completely before birth. In severe cases, the spinal cord can bulge out of an infant's lower back. This protrusion may be as large as a grapefruit, and it may or may not be covered with skin. Spina bifida is the most common *neural tube defect* (NTD). The *neural tube* is the part of a fetus that eventually develops into the brain and spinal column. As many as one out of every 1,000 babies is born with spina bifida in the United States each year.

There are three types of spina bifida, ranging from mild to severe. Children born with the most severe type of spina bifida, known as *myelomeningocele* (mye-uh-loh-meh-NIN-goh-seel) *spina bifida,* may also be born with hydrocephalus. *Hydrocephalus* (hye-droh-SEFF-uh-luss) is a condition in which fluid builds up around the brain. Spina bifida can lead to bladder and bowel control problems, paralysis, and foot and leg problems. In some cases, spina bifida can also result in learning disabilities.

Fast Fact

The term *spina bifida* comes from Latin words meaning "spine split in two."

The exact cause of spina bifida is yet unknown. Some studies have pointed out a link between spina bifida and a lack of folic acid in the diet of pregnant women. *Folic acid* is a B vitamin that is found in fresh fruit, eggs, orange juice, dark green leafy vegetables, and dried beans and peas. According to the March of Dimes, if all women got enough folic acid before and during early pregnancy, the number of babies born with spina bifida might be reduced by 70 percent.

Spina bifida usually occurs during the first month of fetal development. Because this birth defect occurs at such an early stage of pregnancy, many women don't even know that they are pregnant until after the baby has already developed spina bifida. For this reason, it is especially important for women who are thinking of becoming pregnant to eat a healthy diet that contains foods high in folic acid.

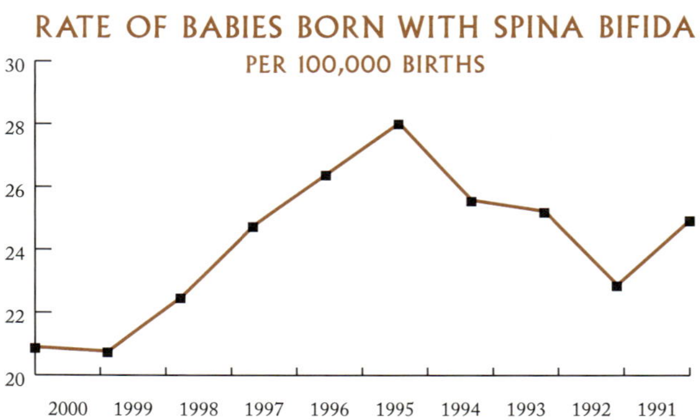

Once a baby has been born with spina bifida, treatment depends on how severe the condition is. In some cases, doctors may operate almost immediately after a baby's birth to push the spinal cord back through the vertebrae and close the hole. Later, other operations may be necessary to correct some of the other problems associated with this condition.

Sometimes, spina bifida is detected *in utero,* or when the baby is still in the womb. Special experimental surgery has been performed to repair cases of severe spina bifida before the baby is even born. The results have shown promise, but it is too soon to tell whether this surgery will become a common treatment.

OTHER NUTRITIONAL DEFICIENCIES

Beriberi (berr-ee-BERR-ee) can be caused by a lack of *thiamine,* a B vitamin, in the diet. It affects the muscles, causing weakness and paralysis. It is most often seen in areas where people have a lot of rice in their diets. White rice lacks the thiamine found in the husks of brown rice.

Osteoporosis can be worsened by a lack of calcium in the diet. People who do not get enough calcium may have thinner, more fragile bones. This, in turn, leads to fractures and other problems.

Tooth decay is caused by a lack of fluoride. In the 1930s, scientists discovered that people who lived in areas where the water was naturally fluoridated had less tooth decay. Half of all Americans now drink fluoridated water. This has reduced tooth decay by 65 percent.

GOUT

Some conditions are caused by diet deficiencies. *Gout,* however, is sometimes caused by a diet that is too rich. This disorder occurs when deposits of *uric acid* crystals collect in the joints and connective tissues. The sharp crystals cause swelling and pain, especially in the big toes. Doctors diagnose gout by checking for uric acid crystals in joint fluid.

Gout may occur when the liver secretes more uric acid than the body can handle. Gout may also be caused by a diet that contains too many rich foods. In such a case, the patient must alter his or her eating habits. Patients with gout are told to avoid red meat, cream sauces, alcoholic beverages, and junk food. Fresh fruits and vegetables are recommended—especially cherries which reduce the level of uric acid in the body.

New York Yankees pitcher David Wells has a hole cut in the front of his right shoe to relieve the pain caused by gout. Wells pitched a no-hitter against the Minnesota Twins in 1998.

BAD BACTERIA

Bacteria are microscopic, single-celled organisms. Bacteria are everywhere. They're floating in the air that we breathe and swimming in the water that we drink. They live on our skin and inside our bodies.

There are thousands of different types of bacteria, but only a few cause diseases and other problems. Some bacteria can actually be beneficial to human health. Doctors have recently discovered that when children are exposed to certain types of bacteria, they are less prone to allergies and skin rashes.

"Bad bacteria," however, can cause serious illnesses and long-lasting health problems. They may cause muscle and joint pain, fever, diarrhea, and other conditions. Diseases caused by bad bacteria include cholera, dengue fever, typhoid fever, and leprosy.

How are humans exposed to these bad bacteria? Some people eat contaminated food or drink contaminated water. Others become ill after being bitten by mosquitoes or other creatures that are carrying the bacteria.

Some types of bad bacteria that affect the musculoskeletal system are transmitted to humans through the bites of ticks. *Ticks* are parasitic mites that live by feeding on the blood of other animals. There are hundreds of different types of ticks around the world. Although not all of these ticks carry harmful bacteria, some do: About thirty major diseases can be transmitted by tick bites.

LYME DISEASE

In the early 1970s, doctors and researchers began wondering why many of children in and around Lyme, Connecticut, were developing arthritis. They realized that all the children lived near wooded areas that were filled with ticks. They also noticed that the children's symptoms started in the summer months, at the height of tick season.

Around 1975, researchers discovered that the origin of *Lyme disease,* which can cause arthritis, was bacteria transmitted through the bites of deer ticks. A deer tick is about the size of the period at the end of this sentence. The ticks themselves became infected after sucking the blood of mice and rats that carried the bacteria.

Today, Lyme disease is the most common tick-borne disease. It affects both young and old, male and female alike. Cases have been documented in nearly every U.S. state. Lyme disease has also been found in Europe and Asia.

For many people, the first sign of Lyme disease is a red, ring-shaped rash that may appear near the tick bite. The rash may be as small as a dime or large enough to cover a person's entire back. As the bacteria spread throughout the body, rashes may appear in other location besides the bite site. People infected with the bacteria may develop flu-like symptoms, including fever, headaches, stiff necks, achy joints, and fatigue.

Dr. James Oliver looks at one of the specimens in his tick collection at Georgia Southern University. In addition to Lyme disease, scientists have discovered at least seven other illnesses that are caused by tick bites.

If Lyme disease is diagnosed early, it can be easily cured with antibiotics. *Antibiotics* are medicines that kill bacteria. However, if the bacteria are allowed to remain in the body for a few months, then they can cause *infectious arthritis.* Infectious arthritis, caused by exposure to bacteria or viruses, affects the joints like other forms of arthritis. Lyme disease can also affect the body's nervous system, causing facial paralysis, confusion, and meningitis (men-in-JYE-tiss), which is inflammation of the covering of the brain and spinal cord.

Tick Protection

If you live near or have ever visited the country, chances are that you've seen or even been bitten by a tick. Here's how to minimize your chance of tick bites while hiking or walking in the woods.

- Stick to well-marked trails and avoid overhanging grass or brush that may harbor ticks.
- Wear long-sleeved shirts and long pants that fit snugly at the wrists and ankles. Tuck pants into socks and always wear a hat.
- Light-colored clothing makes it easier to spot ticks.
- After hiking, inspect yourself carefully to make sure that there are no ticks on you or your clothes. Pay careful attention to hairy areas. Remember: Deer ticks are tiny and may look like freckles or specks of dust.
- Check your pets for ticks. Not only can pets develop Lyme disease, but the ticks that they carry into the house may eventually wind up on *you!*
- If you discover a tick on yourself, ask an adult to help you remove it. The best way to remove a tick is to use tweezers to carefully and gently pull the entire tick out of your skin. Make sure not to squeeze the tick's body. After the tick is removed, wash the bite area with an antiseptic such as alcohol.

ROCKY MOUNTAIN SPOTTED FEVER

Rocky Mountain spotted fever (RMSF) is another tick-borne condition that affects the musculoskeletal system. RMSF is characterized by headaches, fever, and aching bones and muscles. A rash follows one to ten days after other symptoms appear. The rash may eventually change to look like bruising or blood beneath the skin. In severe cases, there may be damage to the liver, kidneys, and heart. In rare cases, death may result. The risk of death from RMSF is higher for people over seventy.

The bacteria that cause RMSF are carried by wood ticks in the West and dog ticks in the Northeast. Nearly every U.S. state has reported cases of RMSF. As with Lyme disease, the tick carries bacteria that cause the sickness when transmitted to humans by a bite. Doctors treat RMSF with antibiotics and rest. Most patients make a full recovery with no lasting effects.

Lyme disease and RMSF are just two of the many conditions that are caused by the bites of infected ticks. Other musculoskeletal conditions caused by ticks include *ehrlichiosis* (ur-lik-ee-OH-siss), *tick paralysis, tularemia* (too-luh-REE-mee-uh), and *Colorado tick fever.* For more information on these conditions, check a medical dictionary.

Tick Traps

One recent innovation in the fight against Lyme disease is a mousetrap that may help kill ticks that carry the Lyme bacteria. Mice are lured into a childproof, pet-proof box. As they pass through the box, their coat is covered with an oily substance that kills ticks. In tests, use of the box has reduced the number of ticks on a 10-acre (4-hectare) plot of land by 96 percent. Further tests are being performed.

OSTEOMYELITIS

Osteomyelitis (ah-stee-oh-mye-uh-LYE-tiss) is a bone infection that may be caused by bacteria or fungi in the bloodstream. *Fungi* are parasitic organisms, such as mushrooms and molds, that feed on both living and decaying matter. The bacteria that is most responsible for osteomyelitis is *staphylococcus* (staff-ih-loh-KAHK-uss), or "staph." Infections caused by the staph bacteria are called *staph infections.* Staph bacteria usually live on the skin of humans without causing problems or illness. When the bacteria enter the bloodstream through an open wound in the skin, however, they can cause a number of different infections.

In osteomyelitis, once the bacteria have infected the bone, the bone produces *pus* (thick yellowish fluid made up of living and dead white blood cells, dead tissue, and bacteria). This in turn may lead to an *abscess,* a collection of pus surrounded by inflamed tissue. Bone tissue is shut off from its blood supply and gradually dies. Those suffering from osteomyelitis may feel pain in the affected bones. They may also run a fever and feel nauseous and dizzy. Osteomyelitis is treated with antibiotics. In chronic cases, surgery may be needed to remove dead bone tissue. In some severe cases that are resistant to antibiotics, amputation of the affected area is necessary.

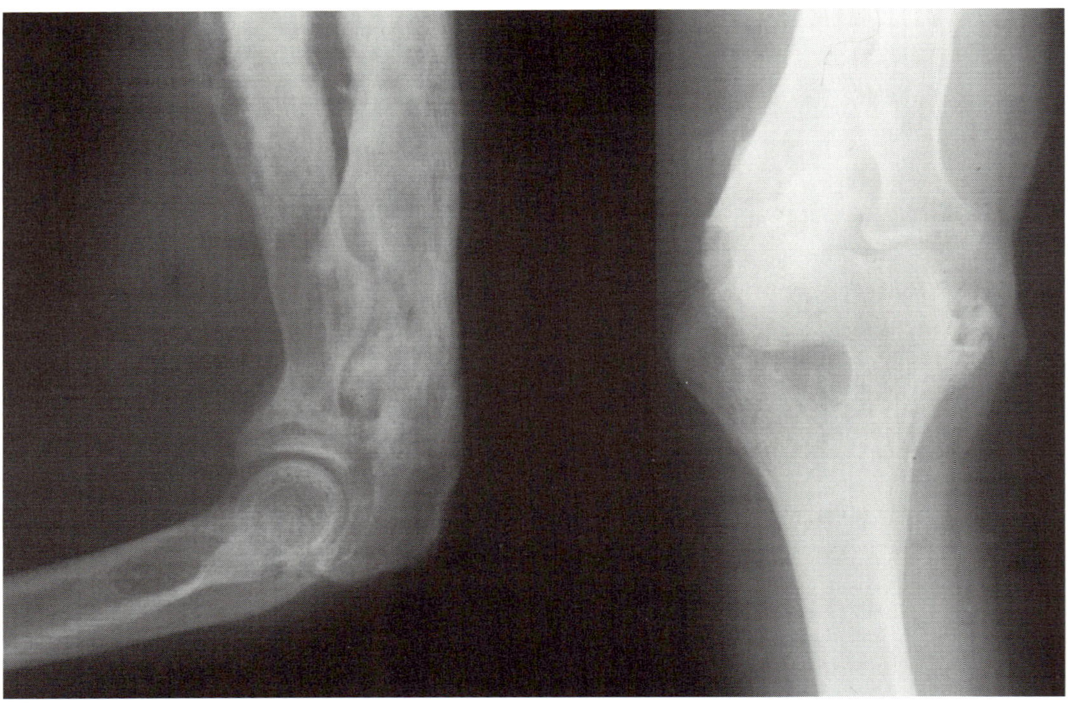

This X-ray shows healthy elbow bones on the left and bones affected by osteomyelitis on the right.

SINUSITIS

Sinusitis (sye-nuh-SYE-tiss) occurs when the cavities in the bones behind the cheeks, eyebrows, and jaw become infected. These cavities are called *sinuses.* Symptoms of sinusitis, which is caused by bacteria or fungi in the body, are inflammation, pain in the sinuses, runny nose, greenish phlegm, fever, fatigue, and sore throat.

Sinusitis might start off as a cold. When the body's defenses are weakened, bacteria that live in the upper respiratory (RESS-pur-uh-tor-ee) tract (the nose, nasal passages, and upper pharynx) move into the sinuses, causing problems there. One such bacterium is *Streptococcus pneumoniae* (strep-toh-KAHK-uss noo-MOHN-yay). Doctors usually prescribe antibiotics to fight the infection.

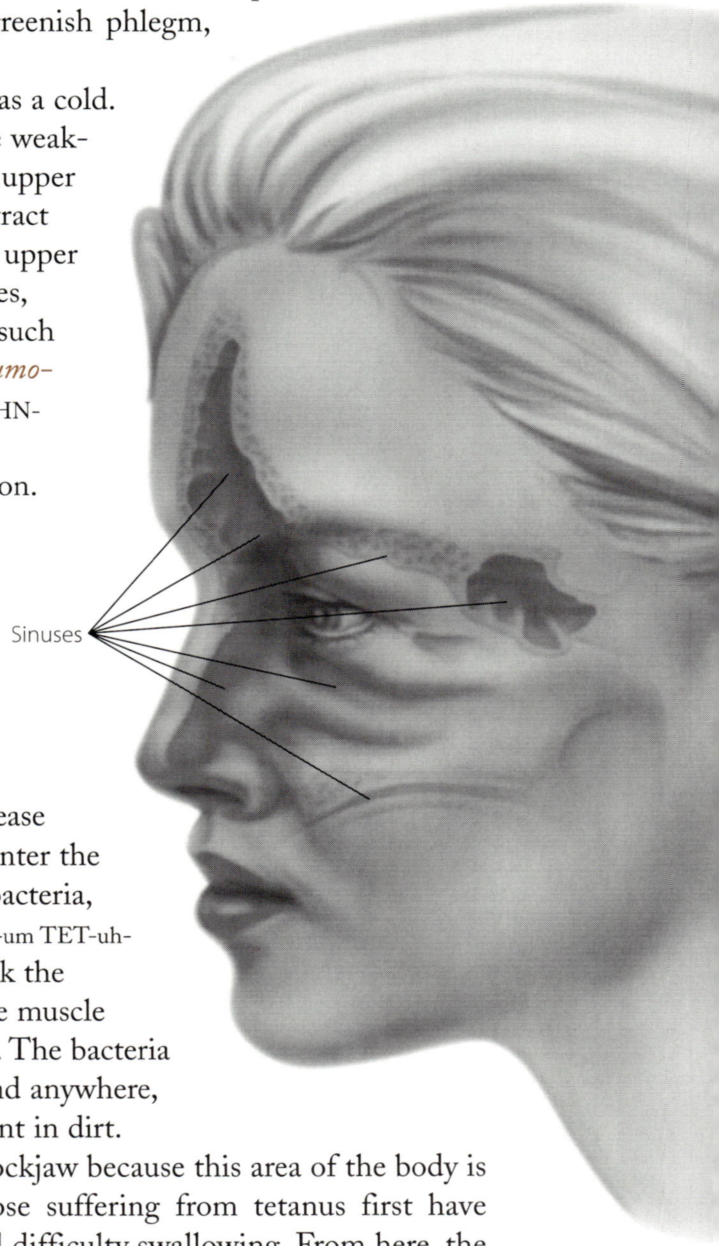

TETANUS

Tetanus (TET-nuss) is a disease that is caused when bacteria enter the body through a wound. The bacteria, *Clostridium tetani* (klos-TRID-ee-um TET-uh-nye), invade the body and attack the nervous system, causing severe muscle contractions that do not relax. The bacteria that cause tetanus can be found anywhere, but they are especially prevalent in dirt.

Tetanus is often called lockjaw because this area of the body is the first to be affected. Those suffering from tetanus first have headaches, then sore jaws and difficulty swallowing. From here, the bacteria can affect muscles throughout the body. Many tetanus cases are fatal, but the condition can be successfully treated with antibiotics or surgery if diagnosed early enough.

Today, vaccines are available to prevent people from getting tetanus. A *vaccine* is a medicine that contains dead or weakened germs. These are introduced into the body to produce antibodies, which attack any live germs and remain in the blood to fight similar germs that may come later. Doctors recommend that all children receive vaccination shots to ward off tetanus. They also recommend that pregnant women receive tetanus booster shots in order to prevent their newborns from getting the disease.

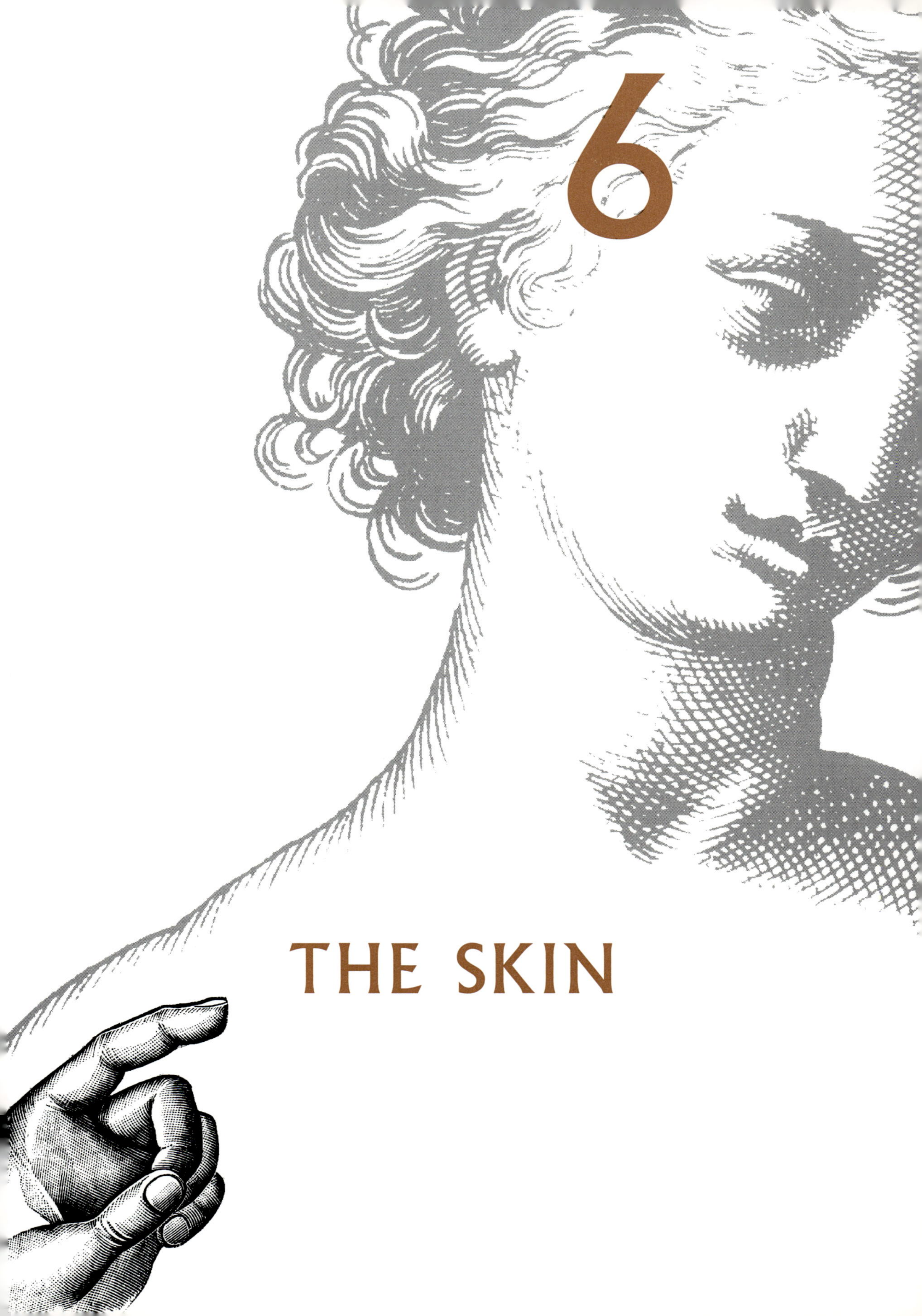

6

THE SKIN

The skin is an amazing, elastic organ that performs a number of very important functions. Not only does the skin cover and protect the entire outer surface of the human body from contaminants, it also regulates body temperature, acts as a sensing organ, and can even repair itself.

The skin is constantly exposed to the environment in which we live. It can be harmed by physical injuries, bacteria, and chemical and natural agents. It is important to care for the skin and protect it from damage. Protective clothing and sunscreens are just two ways that people can protect their skin from harm.

The skin can also be affected by changes or problems inside the body. For this reason, rashes, blisters, and other skin conditions may be a sign of a serious medical condition. Scarlet fever, typhoid fever, and Lyme disease, for example, all have distinctive rashes that help doctors identify them.

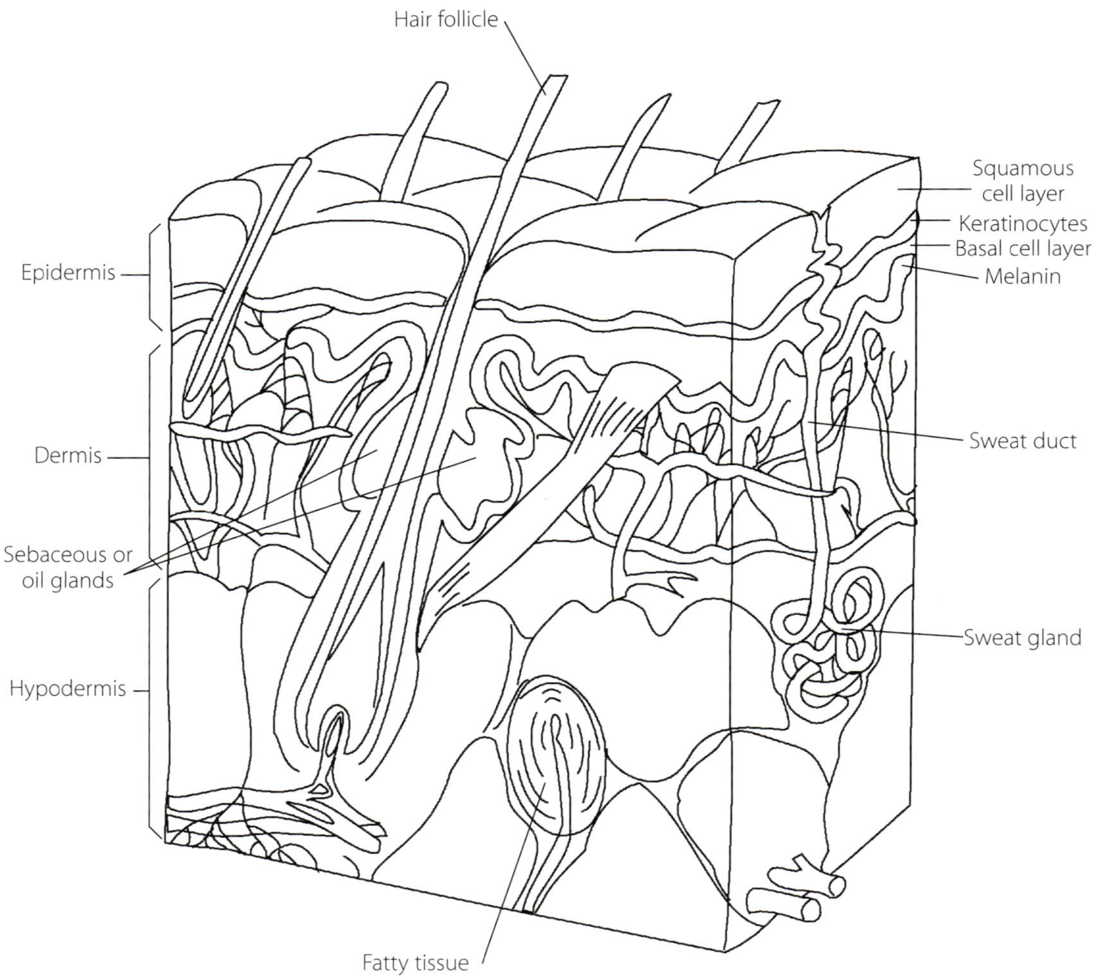

LAYERS OF THE SKIN

The skin has three main layers: the epidermis, dermis, and hypodermis. The *epidermis* is the skin's paper-thin outer layer. It is mostly made up of dead skin cells. The epidermis is the first line of the body's defense system against the environment. It keeps bacteria, viruses, and other pollutants out of the body while keeping in water and important nutrients that the body needs.

The exact thickness of the epidermis is not the same on all parts of the body. The epidermis is thickest, for example, on the soles of the feet. Here, the skin layer measures about 0.17 inches (4.3 millimeters). The epidermis is thinnest on the eyelids, where it measures about 0.02 inches (0.5 millimeters).

The deepest segment of the epidermis is called the *basal cell layer.* Here, cells divide quickly, pushing new cells upward to replace dead cells. The uppermost section of the epidermis is called the *squamous cell layer.* As the cells rise up toward the squamous cell layer, they flatten, shrink, and die. Dead skin cells, called *keratinocytes* (keh-RAH-tih-noh-sites), are made up mostly of a protein called *keratin* (KEH-ruh-tin). A skin cell's journey from the basal cell layer to the squamous cell layer takes about four weeks.

Pigment in the epidermis, called *melanin* (MEL-uh-nin), determines each person's skin color. People with more melanin have darker skin tones, while those with less melanin have lighter skin tones. Skin tone is determined by genetics. It may be altered by exposure to sunlight or as a result of certain skin conditions.

The *dermis,* also called the "true skin," is the middle skin layer. The dermis is made up of water, proteins, blood vessels, nerve endings, and fat cells. In addition, the dermis contains three specialized structures that are found nowhere else on the body: hair follicles, oil glands, and sweat glands.

Hair follicles (FAHL-ih-kulz) produce hair. These follicles are found everywhere on the body except the palms of the hands and the soles of the feet. Some of the hair produced by the follicles is thick and coarse, like the hair on people's heads. Most of it, however, is fine and light.

Oil glands, also called sebaceous glands, are attached to hair follicles. Oil produced by these glands protects the skin from damage. There are more oil glands in some parts of the body than in others. The dermis of the face and back, for example, contains more oil glands than other areas.

The *sweat glands* of the dermis play a key role in regulating the body's temperature. When the body becomes too hot, sweat glands release moisture, or sweat. The sweat evaporates on the skin surface, causing the body to cool off. Humans have millions of sweat glands located over the entire body.

The *hypodermis,* also called the subcutaneous layer, is the deepest skin layer. Located below the dermis, it is made of connective tissue that mainly produces fat. The hypodermis not only anchors the skin to the tissues below, but it also provides the body with insulation and protection. The thickness of the hypodermis varies in each person. It also varies depending upon its location in the body. Fat stored in the hypodermis can be used by the body for energy as needed. Vitamin D is also produced in the hypodermis.

Fascinating Skin Facts

- The average person's skin covers about 2 square yards (1.6 square meters) and weighs about 6 pounds (2.7 kilograms), making it the largest organ in the human body.
- The skin is the part of the body that is most directly in contact with the environment.
- One square inch (6.5 square centimeters) of skin contains about 19 million cells, 90 oil glands, more than 600 sweat glands, 65 hairs, and 19 feet (5.7 meters) of blood vessels.
- The parts of the body with the most sweat glands are the forehead, underarms, palms of the hands, and soles of the feet.
- Hair and nails are considered part of the epidermis. They are both made up of keratin, the same protein that makes up dead skin cells in the epidermis.

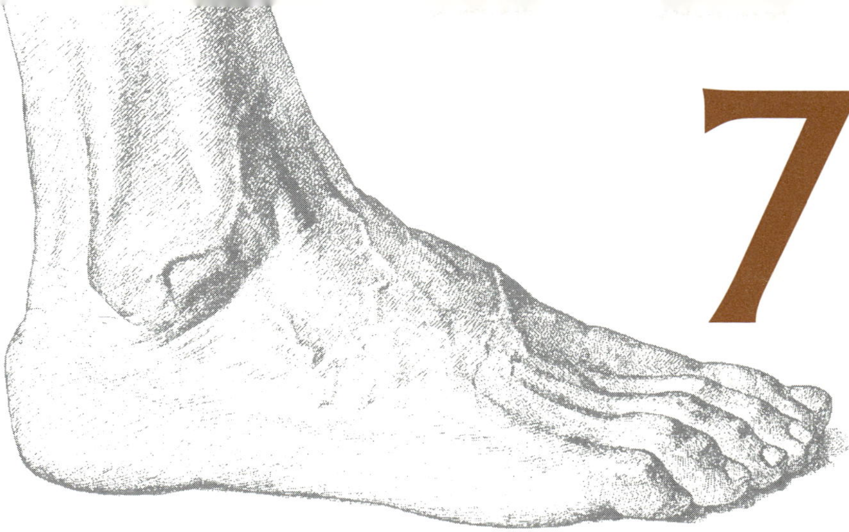

7

SKIN DISORDERS

As with any other part of the body, medical conditions can affect the skin. These conditions may be the result of genetic disorders, autoimmune conditions, or other problems. Here are a few of the many conditions that can affect the skin.

GENETIC DISORDERS

There are a number of genetic skin disorders.

BIRTHMARKS

Birthmarks are congenital (kun-JEN-ih-tul) skin blemishes, meaning that they are present at birth. They result when groups of blood vessels clump together. After birth, they may fade and eventually disappear. Others, however, remain for life. They can be removed by surgery if a person feels that it is necessary. There are two common types of birthmarks.

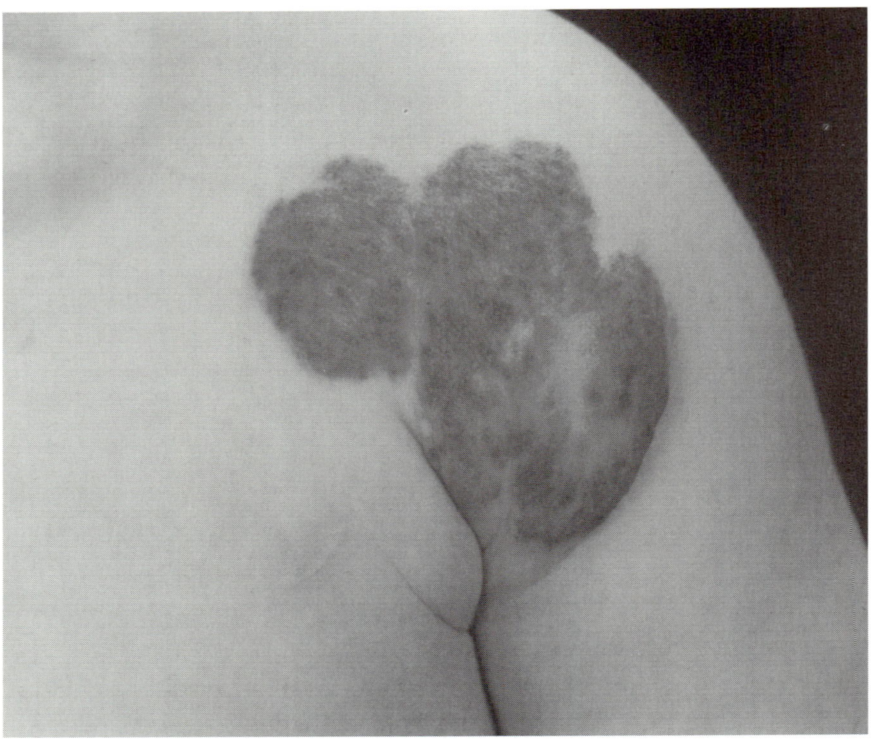

This photograph shows the kind of reddish vascular birthmark sometimes known as a strawberry mark.

Vascular birthmarks are reddish markings that develop before or shortly after birth. Strawberry marks, "stork bites," port-wine marks, and other red-colored markings are vascular birthmarks. Vascular birthmarks are usually harmless.

Pigmented birthmarks are brown, black, or bluish markings that appear before or after birth. Moles (also called nevi) are a type of pigmented birthmark. Most people have from ten to forty moles on their bodies. Moles are usually harmless. However, there is a risk of moles becoming cancerous with overexposure to UV rays.

PIGMENTATION DISORDERS

Pigmentation disorders are conditions that affect a person's melanin production. One example is *xeroderma pigmentosum* (zir-oh-DERM-uh pig-men-TOH-sum), a genetic disorder that makes people extremely sensitive to sunlight. When people with this condition are exposed to sun, they develop dark spots that look like freckles. These spots can eventually turn cancerous. Both the skin and the eyes can suffer serious damage.

Incontinentia pigmenti (in-kahn-tih-NEN-shee-uh pig-MEN-tee) is a rare genetic disorder that results in excessive deposits of melanin. This disorder usually affects children, but then disappears by adulthood. Symptoms include redness and blistering on sections of the skin, followed by wartlike growths and discoloration of the affected areas. Severe cases may result in hair loss, dental problems, vision problems, muscle weakness, and seizures.

Albinism is a genetic disorder in which the skin produces little or no melanin. People with albinism have very pale or white skin and pink or light-colored eyes. Their skin burns very easily in the sun, and they may have vision problems.

Other inherited skin conditions include *Ehlers-Danlos* (ELL-erz DAN-lohse) *syndrome,* which results in loose or stretchy skin that bruises easily; *cutis laxa* (KYOO-tiss LAKS-uh), a condition in which the skin hangs in folds; and *epidermolysis bullosa* (ep-ih-dur-muh-LYE-siss buh-LOH-suh), a congenital disorder that can cause blistered skin from even a minor injury.

For some genetic disorders, doctors can test a baby while it is still in its mother's womb to see if it has the gene that causes one of these conditions. This type of testing is called prenatal testing. In other cases, adults can be tested before conception to see if they have genes that could cause certain conditions if they are passed on.

AUTOIMMUNE DISORDERS

There are a number of autoimmune disorders that can affect the skin.

PSORIASIS

Psoriasis (soh-RYE-uh-siss) is a chronic skin disorder that occurs when cells on the outer layer of skin grow faster than the body can shed them. The skin piles up on the surface, resulting in bright red or pink dry areas with a silvery, scaly surface. Psoriasis, which is believed to be an autoimmune disorder, most commonly affects adults.

Although psoriasis cannot be cured, it can be treated. Doctors use medicated lotions and creams to reduce swelling and discomfort. Psoriasis is also treated with daily exposure to limited amounts of sunshine (without sunburn), also known as *phototherapy.* In severe cases, doctors may prescribe oral medications, as well. If untreated, psoriasis can lead to arthritis in the joints of the fingers and even the spine.

SCLERODERMA

Scleroderma (sklur-uh-DUR-muh) is actually a form of arthritis that can affect the skin. This condition may also affect muscles, joints, and connective tissues. *Localized scleroderma* consists of either reddish patches on the skin that gradually develop into oval-shaped, ivory-centered patches; or long bands that run down the arms, legs, or forehead. Localized scleroderma can affect the skin of the chest, stomach, back, face, arms, and legs. Although this type of scleroderma usually disappears after three to five years, people are sometimes left with darkened areas of skin and weakened muscles.

Systemic scleroderma is much more serious. In addition to the skin, it can also spread to affect tissue, blood vessels, and internal organs. One symptom of systemic scleroderma is calcium deposits forming in the connective tissue. These deposits may break through the skin and cause *ulcers,* or open sores. Other symptoms include a lack of circulation to the feet and hands, causing extreme susceptibility to cold weather; chronic heartburn, thick, tight skin on the hands, and red spots on the hands and face. Severe scleroderma can affect the heart, lungs, and kidneys, causing serious health problems.

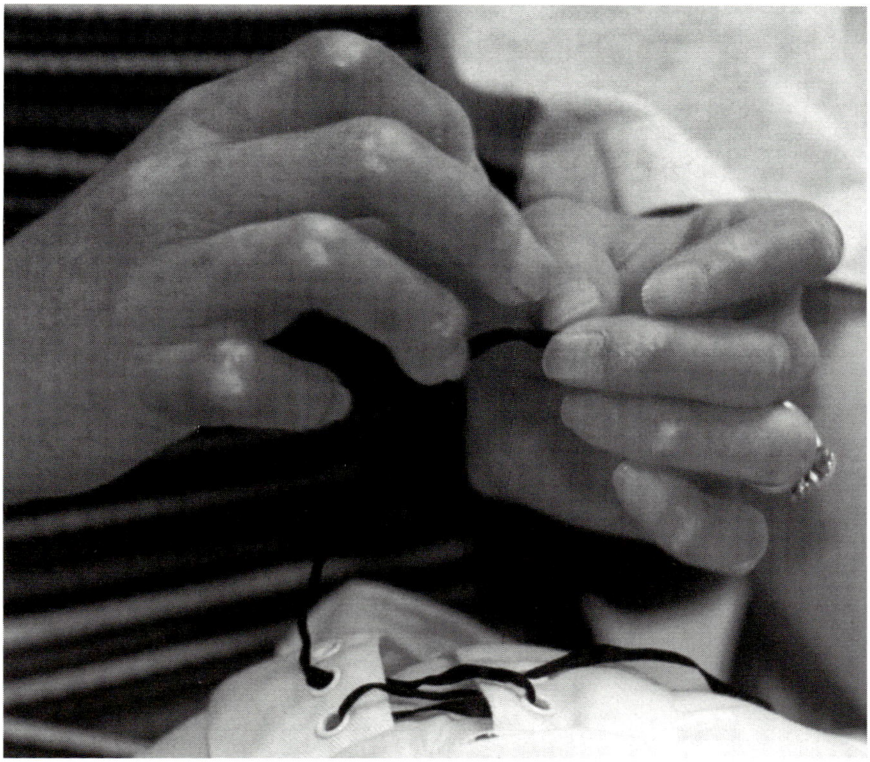

This woman's hands are permanently bent at the joints and her skin has lost most of its elasticity due to scleroderma. The disease causes the immune system to overproduce collagen (scar tissue) that can harden connective tissue in joints and affect internal organs.

Although doctors are not certain what causes scleroderma, it is suspected of being an autoimmune disorder. The immune system is thought to stimulate cells to overproduce collagen. The collagen builds up around connective tissues. There is currently no treatment to stop this from happening, so doctors must treat the symptoms of the condition. People who get heart disorders from scleroderma are treated by cardiologists (kar-dee-AHL-oh-jists), doctors who specialize in heart problems. Patients whose skin is affected are treated by *dermatologists* (dur-muh-TAHL-oh-jists), doctors who specialize in skin care.

Researchers think that environmental factors can trigger scleroderma in some people. For example, viral infections and exposure to certain coating chemicals, as well as vinyl chloride, may trigger an episode of scleroderma.

ALOPECIA AREATA

Alopecia areata (al-oh-PEESH-uh ah-ree-AH-tuh) occurs when the autoimmune system attacks hair follicles inside the skin. This results in patchy hair loss. According to the National Library of Medicine, 2 percent of all Americans will experience alopecia areata at some time in their lives. At this time, there is no effective treatment for this disorder.

MISCELLANEOUS CONDITIONS

There are a number of skin conditions with unknown causes, including the following.

ROSACEA

Rosacea (roh-ZAY-shuh) is a chronic skin condition that causes redness, bumps, and pimples. The most commonly affected areas are the face and upper body. In severe cases, rosacea may cause the skin on the nose to thicken. This condition can also affect the eyes.

About 13 million Americans suffer from rosacea. It generally affects people between the ages of thirty and sixty. People with fair skin are more apt to develop this condition. Although there may be a genetic factor to rosacea, its exact cause is unknown.

One of the first signs of rosacea is a lot of flushing or reddening of the facial skin. This blushing may be triggered by many different things, including spicy foods, hot beverages, alcohol, exercise, and even stress. In some cases, this flushing becomes permanent, leaving the face reddened all the time. In other cases, the nose becomes reddened and swollen.

Some researchers believe that there may be environmental links to rosacea. Tiny mites that live in hair follicles may trigger it. Others believe that there may be a link between intestinal bacteria and rosacea.

Fast Fact

Famous people who have suffered from rosacea include former U.S. president Bill Clinton (1946–), comedian W.C. Fields (1880–1946), and the painter Rembrandt (1606–1669).

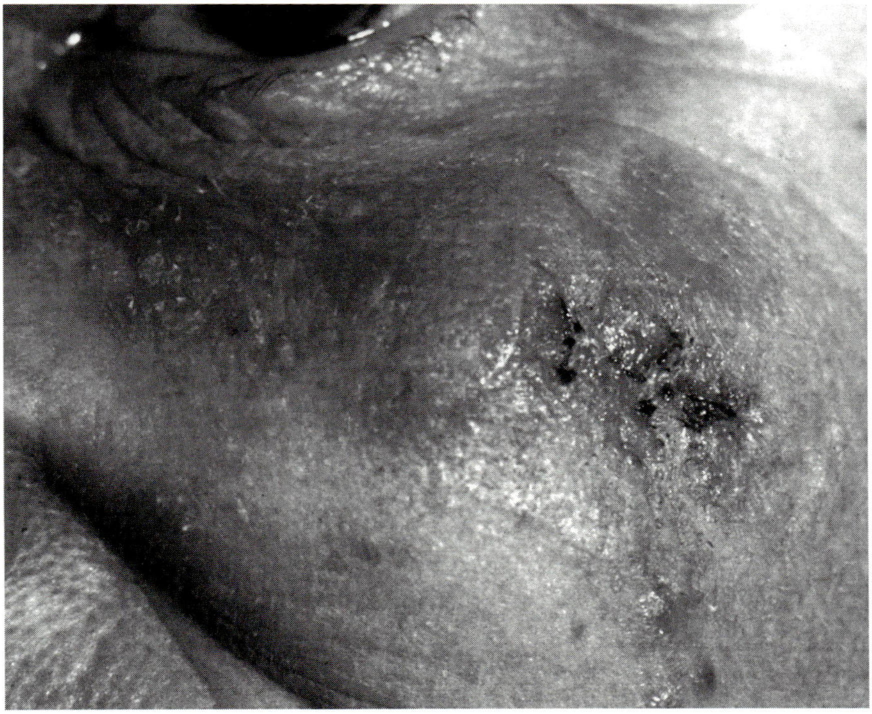

Rosacea often causes redness and bumps on the face. Severe cases can permanently affect the skin and damage the eyes and nose.

Although rosacea cannot be cured, it can be treated and controlled. Doctors may prescribe topical or oral antibiotics to treat rosacea. They also recommend that people affected by rosacea avoid the things that trigger flare-ups. Using sunscreen while in the sun and avoiding facial cleansers that contain alcohol can also help control the condition.

ACNE

Acne is a skin condition in which pores become plugged and infected. Boils, pustules, and pimples result. Acne is the most common skin condition in the United States today. According to NIAMS, nearly 17 million Americans suffer from acne. The condition affects the face, back, neck, chest, and shoulders. Severe cases of acne can result in permanent scarring.

The exact cause of acne is not known. However, the condition may be triggered in teens and preteens by the increase in hormones during puberty. The American Academy of Dermatology reports that nearly 100 percent of people between the ages of twelve and seventeen will get acne. Researchers also believe that some people have inherited genetic factors that make them more likely to suffer from acne.

Dermatologists recommend treating mild cases of acne with an over-the-counter acne medication or wash. More serious cases may require using a stronger medication prescribed by a doctor.

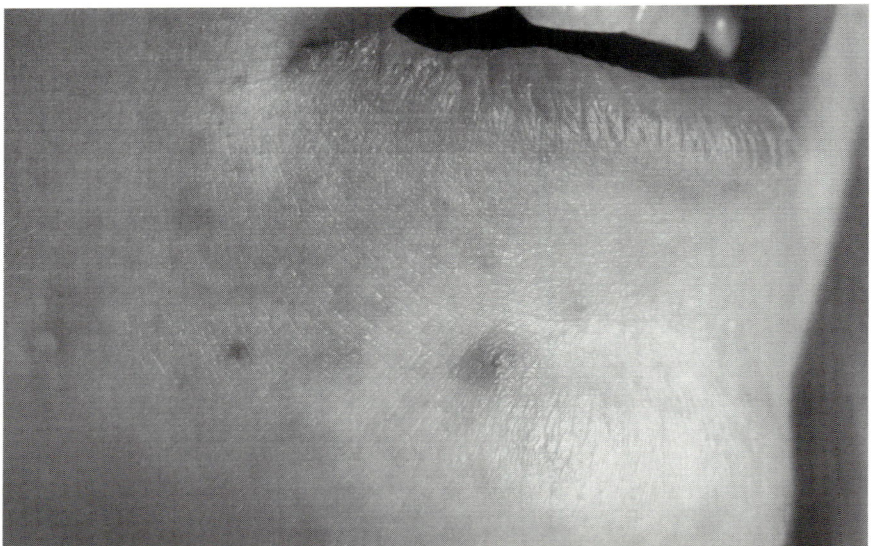

Fascinating Acne Facts

- Dermatologists do not believe that chocolate, french fries, pizza, or other foods affect or worsen acne.
- Pressure from backpacks, bike helmets, and tight collars can cause acne flare-ups.
- Popping, squeezing, or picking at *comedones,* blackheads or whiteheads, can make the condition much worse. It can also cause scarring and pitting of the skin.
- Poor hygiene and dirty skin do not cause acne.
- Doctors recommend that people with acne wash their face gently, using a mild soap, twice a day. Scrubbing too hard or too often may only make things worse.

SEBORRHEA

Seborrhea (seb-or-EE-uh) is a skin condition that results in dry, itchy skin accompanied by white or yellow scales. The condition usually affects the skin of the head, face, and upper body. Seborrhea may be genetic, but it might also be aggravated by other factors, including stress, weather, insufficient shampooing, and even *obesity.* Obesity is the condition of having too much body fat. When infants have seborrhea of the scalp, it is called cradle cap.

VITILIGO

Vitiligo (vih-tih-LYE-goh) is a condition in which melanocytes (MEL-uh-noh-sites), cells that make pigment, are destroyed in the skin, the retinas of the eyes, and the mucous membranes. *Mucous membranes* are the layers of mucus-secreting cells that line the tubes and cavities of the body. The condition results in white patches on the skin in affected areas. Any hair on the affected area also turns white.

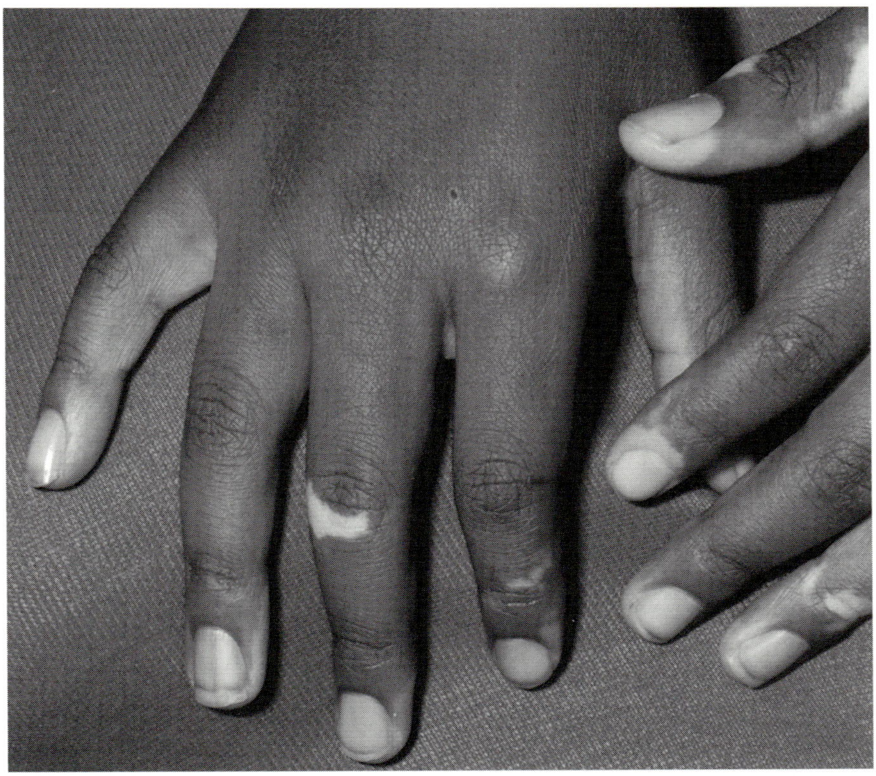

The white spots on this person's hands are a result of loss of skin pigmentation caused by vitiligo.

No one knows what causes this condition, but it affects about 40 to 50 million people around the world. Most people develop the condition before the age of forty. People suffering from vitiligo may be worried, upset, or depressed by the way they look. Treatment includes medicines that can be applied to the skin or taken orally. Skin grafts may also be performed. A *skin graft* is a piece of skin taken from one part of the body and transplanted to another part.

One famous person with vitiligo may be singer Michael Jackson (1958–). In 1993, when many people wondered if he was bleaching his skin to make it look lighter, Jackson publicly stated that he had a skin condition. Jackson's dermatologist later confirmed that the singer suffers from vitiligo. Although this condition most often leaves white patches, Jackson may use makeup to make his skin look one color.

8

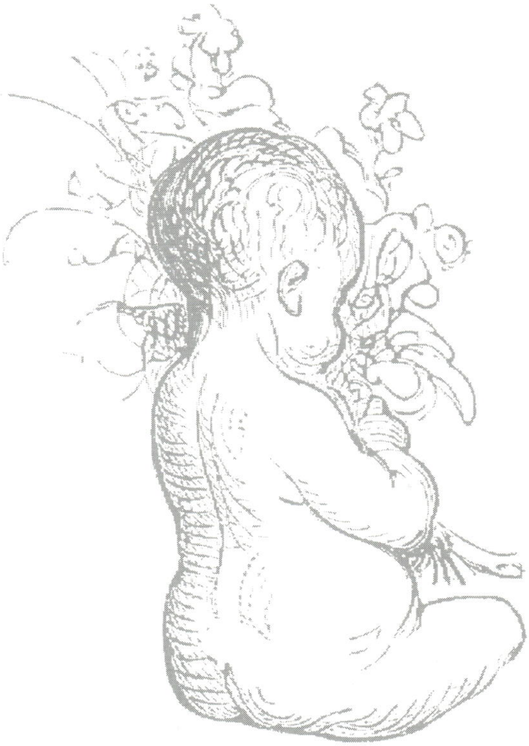

HOW THE ENVIRONMENT AFFECTS THE SKIN

The skin is the first line of defense between the human body and the world around it. Each day, it comes in contact with many substances that can cause irritations, infections, or disease. These irritants can come from a variety of sources.

POLLUTION IN THE WATER

Water pollution can cause big problems for the skin. Pollutants that enter our water include chemicals and other industrial waste, sewage, agricultural runoff from fertilizers and pesticides, and many other substances. In recent years, superpolluted water has caused health officials to close down beaches where people could come in contact with harmful substances. In 2001, 27 percent of all beaches being monitored by the EPA were affected by at least one or more closings or advisories. The majority of beach closings were related to contamination from chemical pollutants and human and animal waste.

As more and more people settle along the world's coastal areas, pollution also increases. Each day, billions of gallons of water, both treated and untreated, are dumped into the ocean. Runoff from streams and rivers also carries pesticides and other harmful chemicals into the ocean.

What happens when people come in contact with pollutants in the water? The first part of the body to be affected is the skin. Contaminated water can cause skin irritations and infections. If there is a cut in a person's skin, toxins, bacteria, and other contaminants can be carried throughout the body.

The skin is not the only part of the body affected by polluted water. Ear, nose, throat, and digestive problems are also associated with water pollution. Scientists believe that those most at risk are children and people with damaged immune systems. Those who spend a lot of time in the water may also be at a higher risk.

Not all skin irritations in the water result from pollution problems. "Swimmer's itch," for example, is an infection caused by a parasite that penetrates the skin. Another irritation caused by living sea creatures is "seabather's eruption," also known as sea lice. Swimmers experience itchy rashes when jellyfish or sea anemone larvae become trapped in their bathing suits. The larvae secrete toxins that cause the irritation.

A boy with open sores on his back watches his friends play in the Yamuna River near Delhi, India. Although the Yamuna is Delhi's main source of water, it is so polluted that the water is black.

Chlorine and Skin Irritation

Not all the chemicals in our water are pollutants. Some are added intentionally. In many communities, for example, the chemical *chlorine* is added to drinking water and swimming pool water in order to destroy bacteria.

There are many health benefits to be had from chlorine. However, some researchers believe that chlorinated water can cause premature aging. It may also increase the risk of melanoma (mel-uh-NOH-muh), or skin cancer.

POLLUTION IN THE AIR

It's hard to escape air pollution—it's all around us. While breathing problems are the most obvious results of air pollution, pollutants in the atmosphere can also cause some nasty skin conditions. Here are some pollutants and the skin problems that they can create.

Residents of Hong Kong cover their mouths and noses to reduce the effects of breathing in smog caused by industrial pollution and heavy automobile traffic. Air pollution continues to be a major health threat in major cities throughout the world.

Health and History: PCBs and Disease

In 1968, more than 1,000 people in Japan became ill after eating food cooked in PCB-contaminated rice oil. The condition was named *Yusho,* or "oil disease." People who had ingested the contaminated oil suffered severe acne and skin sores, muscle aches, numbness, and headaches. In many cases, the acne lingered for years and left behind scars. The PCBs also affected the unborn. Some women who had eaten the poisoned food gave birth to babies with various kinds of birth defects and health problems.

- *PCBs* can cause acnelike skin rashes in adults and more serious neurological problems in children. People ingest PCBs when eating fish or wildlife contaminated with these chemicals. In addition, old appliances, including refrigerators and electrical equipment, may contain PCBs. It is best to avoid sites contaminated with these chemicals.
- *Agent Orange,* in addition to causing cancer, may also cause skin conditions, including chloracne. Chloracne (klor-AK-nee) is severe acne characterized by rashes and cysts. Agent Orange may also cause porphyria cutanea tarda (por-FIR-ee-uh kyoo-TAY-nee-uh TAR-duh). People who have this skin disorder experience blisters, thinning of the skin, ulcerations, and other problems when their skin is exposed to the sun.
- A chemical called *formaldehyde* (form-AL-deh-hyde), found in plywood, wood paneling, tobacco smoke, glues, and other sources, may cause skin rashes and other allergic reactions. Formaldehyde can also cause eye irritation and breathing problems.

Arsenic and Skin Cancer

In some places around the world, people are exposed to the chemical arsenic, which is a known human *carcinogen* (kar-SIN-uh-jen), or cause of cancer. Arsenic can be ingested when people drink water, eat food, or breathe air that has been contaminated with this potentially deadly toxin.

In a part of Taiwan, some wells are contaminated with high levels of arsenic. Research indicates that people in this area are at increased risk for skin cancer. In Chile and Bangladesh, warts, corns, and other skin problems, as well as changes in skin pigmentation, have been noted in people who have been exposed to arsenic in water. In some cases, the corns became cancerous. Arsenic has also been connected with lung, kidney, and bladder cancers.

DERMATITIS

Dermatitis (dur-muh-TYE-tiss) is any inflammation of the skin. It results in swollen, irritated, itchy skin. Dermatitis may be caused by an allergic reaction, excessive scratching, parasitic infection, and many other agents.

One particular type of dermatitis is *atopic* (ay-TAH-pik) *dermatitis,* sometimes known as eczema (EK-seh-muh). It is a chronic skin condition that is characterized by dry, itchy skin. People affected by this condition may also have an angry red rash, hives, or even pus-filled boils. Atopic dermatitis is quite common in children under the age of five. Older children and adults, however, can also suffer from it.

The tendency to get atopic dermatitis may run in some families. The condition itself is usually triggered by an allergic reaction to something in the environment. Triggers include chemicals, detergents, perfumes, medicines, dyes, and natural agents like poison ivy or poison sumac. Researchers believe that people who live in humid, urban areas are more likely to get atopic dermatitis.

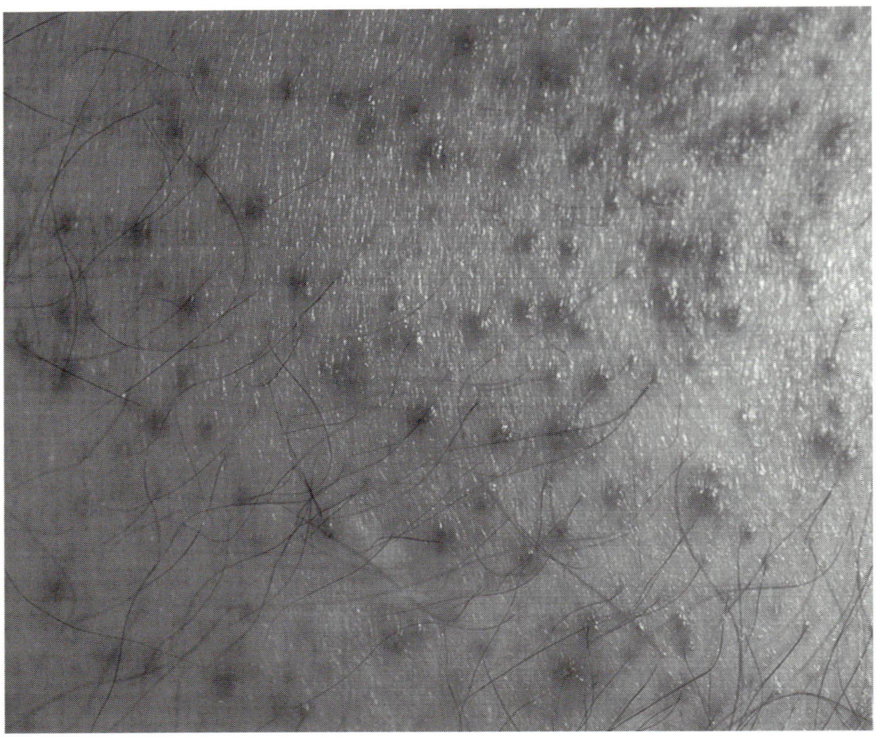

The skin in this photo is inflamed as a result of dermatitis.

People with atopic dermatitis should avoid scratching their skin at all costs. Scratching not only inflames the skin and makes the condition worse, but it may break the skin. This allows bacteria to enter and infect the rash area. Those who find the itching unbearable may need a cortisone lotion for their skin or an oral medication.

Dealing with Dry Skin

How can people with itchy skin avoid scratching?

- Keep your fingernails short.
- Take short baths or showers in lukewarm water. There are some soap substitutes and bath oils containing soothing ingredients that can be added to ease the itching.
- Drink lots of water. This helps to keep skin moist.
- Wear loose clothing.

INFECTIONS AND INFESTATIONS

Some skin conditions are brought on by infections caused by bacteria and other organisms. Others may be caused by infestations of mites or parasites.

SCABIES

Scabies (SKAY-beez) is a contagious skin infection that is caused when tiny mites burrow into the skin to lay their eggs. Symptoms of scabies include small blisters that are very itchy. The arms and hands are most often affected.

Scabies is spread from one person to another through close physical contact and sharing bedding or clothing. The mites that cause scabies can live in clothes, bedding, and dust for up to three days. Children under the age of fifteen are most likely to get scabies. Scabies is treated with a medicated lotion or cream rubbed on the affected areas.

HEAD LICE

Head lice are tiny parasites that live on the scalp. Lice bites can cause inflammation and itching. The lice lay their eggs, called *nits,* in the hair, so the infestation will not go away on its own.

Head lice are not a sign of bad personal hygiene. They are spread from one person to another through close physical contact or sharing bedding, clothing, combs, or brushes. Anyone who comes in contact with an infected person is at risk of getting head lice.

People treat head lice with medicated shampoos that kill the lice and loosen nits from the hair. The treatment may need to be repeated in seven to ten days to ensure that all the lice have been killed. Even one louse can start an infestation.

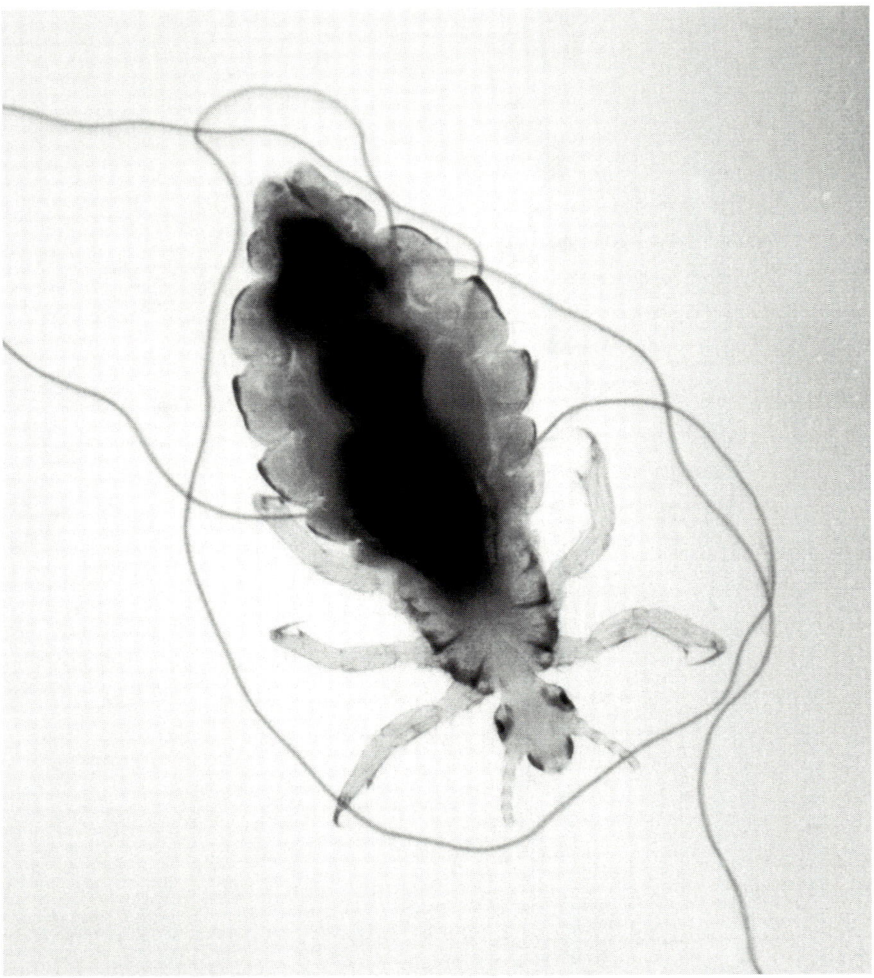

This magnified picture of a human head louse shows its size compared to the human hair that surrounds the parasite.

IMPETIGO

Impetigo (im-puh-TYE-goh) is a contagious skin infection that is caused by bacteria. Symptoms of the condition include fluid-filled blisters and itchy, red skin. Although impetigo can affect any part of the body, it most commonly occurs around the nose and mouth. The bacteria often strike skin that is already suffering from another condition such as poison ivy, scabies, or eczema.

Young children are most often infected with impetigo. It is easily spread from one child to another through physical contact, especially during summer months. Impetigo is treated with antibiotics. Once treatment is started, the rash usually disappears within two or three days.

WEATHER ALERT

Sunshine is an important element for a healthy life. Sunlight is key to the production of vitamin D by the human body. Sunlight can even be used to treat a number of medical conditions. However, too much sunlight can cause some serious health conditions, including sunburn, wrinkles, rashes, and even skin cancer.

> **Fast Fact**
>
> Experts say that by the age of eighteen, most people have already experienced 50 to 80 percent of their total sun exposure.

The rays that cause most of the problems are two types of invisible rays: ultraviolet-A (UVA) and ultraviolet-B (UVB). Even on overcast days, the sun's UV rays still penetrate the clouds and reach Earth—and sunbathers. In response, the skin begins producing melanin to absorb the UV rays and any sun damage. The darker a person's skin, the more melanin he or she has. As melanin increases, the skin grows darker.

The chance of being burned by the sun is greater in summer. When Earth is tilted toward the sun, the sun's rays have less atmosphere to penetrate. However, burns are possible during the winter months, too, especially when people are outside in the snow. The snow acts as a mirror, reflecting the UV rays.

Our Thinning Ozone Layer

The *ozone layer* is an important part of Earth's atmosphere. It helps filter out the harmful UV rays from the sun. However, over the past decades, pollution has deteriorated the condition of the ozone layer. Certain chemicals, including chlorofluorocarbons (klor-oh-FLOR-oh-kar-binz), or CFCs, and halons, have damaged Earth's protective layer.

The ozone layer is thinnest over the Earth's North and South Poles. People who live in countries closest to the Poles must take extra care when out in the sun. South Africa and Australia have the highest rates of skin cancer in the world.

SUNBURN

A *sunburn* is exactly what it sounds like: skin burned by the sun. Sunburns occur when the skin is exposed to a high amount of UV rays. The skin has only so much melanin. People who have pale skin have less melanin, so they burn much more quickly than those with darker skin.

Sunburn can cause severe pain and discomfort. Symptoms include red, hot, swollen skin that is painful to touch, fever, upset stomach, and dizziness. Blisters may form if the burn is a bad one. Because the extent of the burn may not be apparent until hours after the damage has been done, it is not unusual for people to become severely burned without knowing it—until it is too late.

The pain of sunburn can be eased by taking a cool shower or bath. Pain medications and soothing lotions may also ease the burning sensation. For severe burns that are blistering, affected areas should be bandaged with dry gauze. However, the only cure is to wait until the skin heals.

Once the visible effect of a bad burn disappears, the damage may still remain. Just one blistering sunburn doubles the chances that a person will develop malignant melanoma, the most serious form of skin cancer, later in life.

Fast Fact

A 2002 study showed that only one-third of teens surveyed protected themselves with sunscreen when out in the summer sun. Girls were more likely than boys to use sunscreen.

In addition to burns, blisters, and cancer, the sun can cause other skin damage. Premature aging, including wrinkles, brown age spots, and blotchy skin, are some of the effects caused by the sun's UV rays.

Healthy Tan?

Tans are not healthy. A tan indicates that the skin has been damaged. Some people think that they are avoiding skin damage by using tanning beds or tanning lamps to darken their skin. However, the UV rays from a tanning bed are estimated to be five times stronger than the UV rays from the sun.

Always remember that a deep, dark tan is not healthy. If you really want tanned skin, however, experts recommend a tan from a bottle. Some sprays and lotions contain ingredients that react to proteins in the epidermis, darkening the skin. Others contain dyes that add pigment to the epidermis.

SKIN CANCER

Researchers have found a definite link between skin cancer and overexposure to the sun and its UV rays. Studies have shown that 90 percent of all skin cancers develop on areas of the body that are exposed to the sun. The sun's harmful UV rays can cause cancer by damaging the body's DNA.

Skin cancer is the most common type of cancer. According to the National Cancer Institute, 1 million Americans are diagnosed with skin cancer each year, and 40 to 50 percent of all Americans who live to the age of sixty-five will have skin cancer at least once.

Three common types of skin cancer are *basal cell carcinoma* (kar-sih-NOH-muh), *squamous cell carcinoma,* and *melanoma.* Basal and squamous cell carcinomas are easy to treat if caught early enough. Although these types of cancers can be disfiguring, 95 percent of all cases are treatable.

Melanoma, the most common type of cancer for people aged twenty-five to twenty-nine, is more difficult to treat. Melanoma is a type of skin cancer that starts in melanocytes, the skin cells that produce melanin. Melanoma can spread to surrounding tissues and other parts of the body, particularly the lungs and liver.

In addition to sun exposure, a number of other factors may contribute to a person's chances of getting melanoma. People with many moles and people whose family members have a history of the condition may be more at risk. White people, especially those with fair skin and hair and blue, green, or gray eyes, are also more at risk than are darker-complexioned people.

Melanoma accounts for eight out of every ten deaths from skin cancer. If melanoma is diagnosed early, doctors have a better chance of curing a patient. Between 1973 and 1999, melanoma mortality rates in the U.S. more than doubled for women and nearly tripled for men. According to the American Academy of Dermatology, about 7,800 Americans die from melanoma each year.

Doctors treat skin cancer with surgery or radiation treatment. The course of treatment depends on the type of cancer. With melanoma, for example, surgery is almost always required to remove the malignant tumor. If a very large tumor is removed, skin grafts may be necessary to give the area a normal look. If melanoma has spread, chemotherapy may be necessary.

Is It Skin Cancer?

How do doctors distinguish melanoma from other skin blemishes and moles? They look for "ABCD"—*asymmetry, border, color,* and *diameter.*

- **Asymmetry:** A melanoma is often *asymmetrical* (ay-sim-MET-rik-ul). This means that one half of the melanoma is not the same shape as the other half.
- **Border:** A melanoma often has an irregular border. The edges may be ragged, notched, or blurred.
- **Color:** A melanoma may have an uneven mix of colors instead of just one color. Melanomas may be black, bluish, red, or brown.
- **Diameter:** Melanomas are usually larger than 0.25 inches (6.3 millimeters) in diameter. If the tumor has changed in size recently, that is a cause for concern.

Protecting Yourself from Sunburn

How can you protect yourself from a bad burn? Check out these tips.

- Stay out of the sun during the peak hours of ten in the morning to four in the afternoon.
- Use a sunscreen that has a sun protection factor (SPF) of 15 or stronger. Studies suggest that kids who use sunscreen might reduce their risk of getting melanoma later in life by 78 percent.
- Wear sunglasses that filter out UV rays to help prevent damage to the eyes.
- Wear a wide-brimmed hat and protective clothing when in the sun.
- Reapply your sunscreen frequently.

PROTECTING THE SKIN FROM UV RAYS

It's important to protect yourself from the sun year-round, not just in the summertime. To warn people about the potential daily danger of the sun's UV rays, the National Weather Service and the EPA developed a UV index. UV alerts, using the UV index, are regularly included in many weather reports around the country.

What Does the UV Index Mean?

UV INDEX VALUE	PRECAUTION
0–2 (minimal)	Wear a hat or a cap.
3–4 (low)	Wear a hat and apply a sunscreen of SPF 15 or greater.
5–6 (moderate)	Wear a hat, apply a sunscreen of SPF 15 or greater, and stay in shady areas.
7–9 (high)	Wear a hat, apply a sunscreen of SPF 15 or greater, stay in shady areas, and remain indoors from ten in the morning to four in the afternoon.

Experts say that the worst time to be out in the sun is from ten in the morning to four in the afternoon. In the United States, early spring to late summer is prime burning time. However, you know that UV rays can still get you on cloudy or wintry days. Also, the National Institute for Environmental Health recommends that sunglasses be worn at all times in the sun to reduce the risk of cataracts and other eye damage.

Fast Fact

Check your sunscreen's expiration date. Most sunscreens are only effective for a three-year period.

WIND AND COLD

Excessive exposure to windy, cold weather can cause some serious health conditions. Severe cold can cause frostbite and other medical problems, including *hypothermia* (hye-poh-THUR-mee-uh), or abnormally low body temperature, and death.

Frostbite is damage to the skin caused by extreme cold. The skin becomes hard, pale, and cold. Some areas of the body, such as the nose, ears, cheeks, fingers, and toes, are more susceptible to frostbite than others. In some cases, damage from frostbite can extend to the underlying tissues and blood vessels.

Fast Fact

Because it interferes with blood circulation, smoking can increase the risk of frostbite.

People suffering from frostbite should cover the affected areas as soon as possible. However, the frostbitten parts should not be rubbed or chafed.

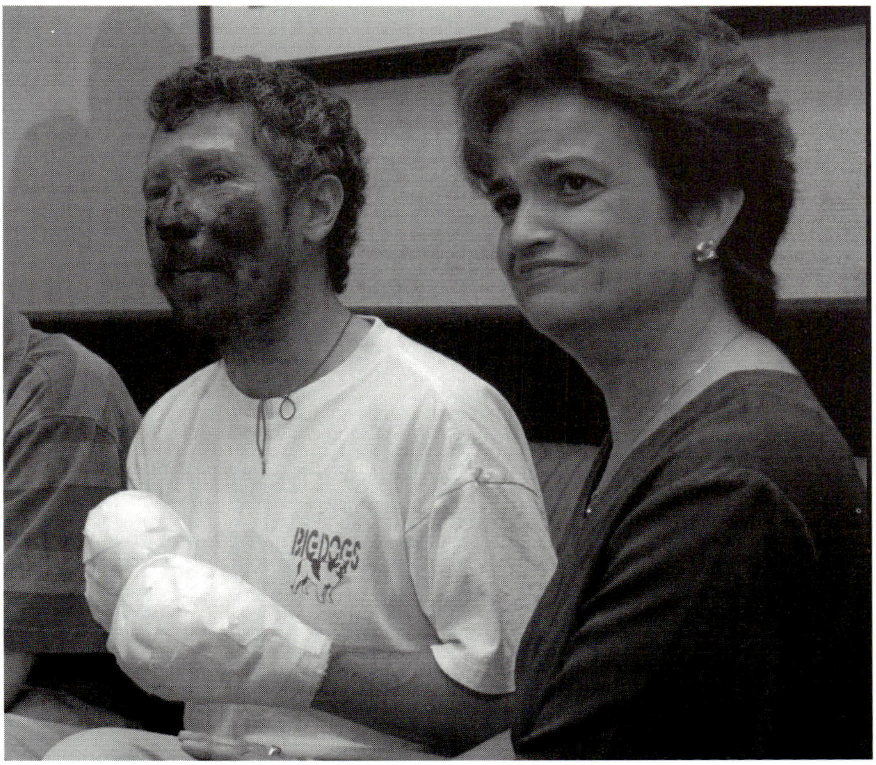

Dr. Seaborn Beck Weathers suffered severe frostbite on a climbing expedition to Mount Everest in 1996. As a result of exposure to the extreme cold, his left hand and part of his right hand were amputated.

Safety If Stranded

Many people risk frostbite when they are unexpectedly stranded in their car in freezing conditions. In such a situation, the CDC recommends the following.

- Remain with your vehicle. Do not try to find help.
- Tie a piece of bright cloth to the antenna to signal rescuers.
- Move anything that you need from the trunk into the passenger area.
- Wrap your entire body, including your head, in extra clothing, blankets, or newspapers.
- Stay awake. You will be less vulnerable to cold-related health problems.
- Run the motor (and heater) for about ten minutes per hour, opening one window slightly to let in air. To reduce the risk of breathing in deadly exhaust fumes, make sure that snow is not blocking the exhaust pipe.
- As you sit, keep moving your arms and legs to improve your circulation and stay warmer.
- Do not eat unmelted snow. It will lower your body temperature.

SPORTS AND THE SKIN

Many athletes experience skin irritations as a result of physical activity. For example, *folliculitis* (fuh-lik-yoo-LYE-tiss) is inflammation of the skin pores. This skin irritation is often caused by wearing wet or sweaty clothing.

Blisters usually occur on areas that are worked hard. Gripping tennis rackets, oars, and other sports gear can cause blisters on the palms. Ill-fitting shoes can cause blisters on the feet. Blisters can be prevented by rubbing the areas with petroleum jelly before exercise. Use foot spray before exercise to reduce perspiration. Also wear acrylic socks and shoes that fit well and are comfortable.

Also called sports-induced acne, *acne mechanica,* an allergic reaction, is caused by wearing tight synthetic jerseys and other sports gear near the skin. Experts recommend wearing clean cotton T-shirts underneath synthetic clothing to prevent this condition.

Tinea (TIN-ee-uh) *infections,* including jock itch, athlete's foot, and ringworm, are caused by a fungus growing on the hair, skin, or nails. Such infections usually have a ring-shaped appearance, with normal skin in the center. These fungal infections are spread from person to person in locker rooms, showers, and other public places. Medicated creams usually kill these infections, although other types of medicines may be needed.

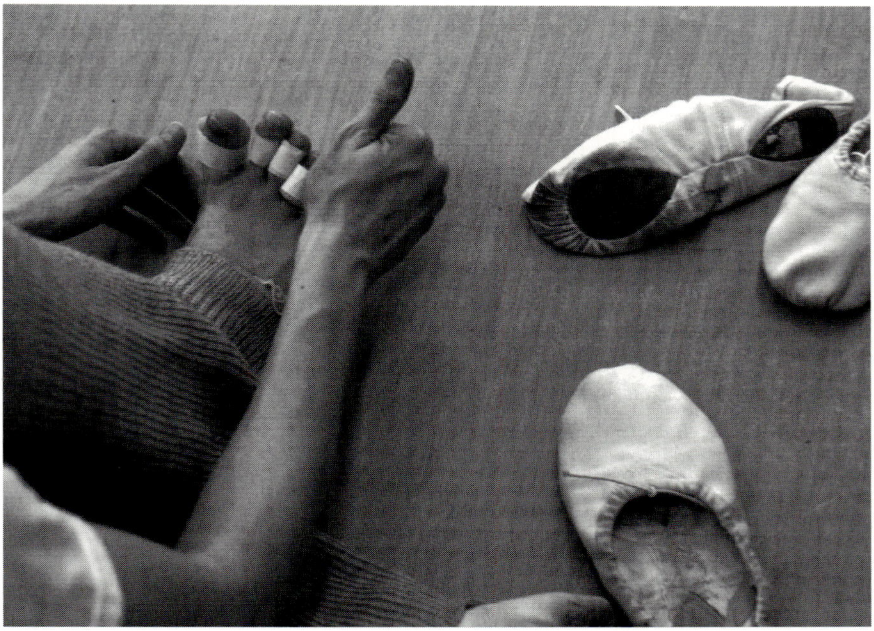

Before a dance rehearsal, a ballerina tapes her toes to prevent blisters.

VIRUSES AND BAD BACTERIA

A number of skin conditions are caused by viral or bacterial infections. Warts, for example, are caused by a virus known as the *human papillomavirus* (pap-ih-LOH-muh-vye-russ), or HPV. There are more than sixty types of HPV. These viruses cause the cells on the outer layer of the skin to grow rapidly. Common warts usually appear on the hands, but can grow on any part of the body. Plantar warts most commonly appear on the soles of the feet. Although some warts disappear on their own, others may need to be removed.

Leprosy, also called Hansen's disease, is an infectious illness caused by bacteria. The bacteria attack the skin and also affect the mucous membranes and nerves. Leprosy can cause loss of skin sensation and muscle paralysis. Serious internal effects include a loss of bone. Up until the late 1800s, there was no cure for leprosy, and people with the condition were shunned and kept away from others. Today, leprosy is much less common than in the past. It is usually found in countries where people do not have access to proper nutrition. Leprosy is very treatable today. People with the condition are given medicine and fed a well-balanced diet.

Childhood Scourges

In recent years, measles, rubella, and chickenpox were common infectious diseases that caused rashes and other more serious symptoms. Today, vaccines can prevent the occurrence of these three once-feared diseases.

- *Rubeola* (roo-bee-OH-luh), more commonly known as measles, is a highly contagious viral disease characterized by small red dots on the skin. Irritated eyes, runny nose, and coughing are also symptoms of the infection. The disease can be fatal if it spreads to the brain. In the United States, rubeola was once one of the most common childhood diseases until a vaccine was developed in 1963. However, it is still common in other parts of the world.
- *Rubella* (roo-BELL-uh), also called German measles, is an infectious viral disease. Less serious than rubeola, rubella shows up as a red rash and can include such symptoms as fever, sore throat, and swollen glands behind the ears. The rash starts on the face and spreads to the lower body. Although usually not serious, rubella can cause birth defects in babies if their mothers contract the disease in the first three months of pregnancy.
- More commonly known as chickenpox, *varicella* (vaar-ih-SELL-uh) was once thought to be unavoidable. This disease, caused by a virus, is spread through the air, by sneezing and coughing. Varicella is well known for its itchy, fluid-filled blisters. The blisters dry up and form scabs in a matter of days. Scratching the blisters may cause permanent scarring. The disease can be serious in people with weakened immune systems. Today, a vaccine exists to prevent varicella.

9

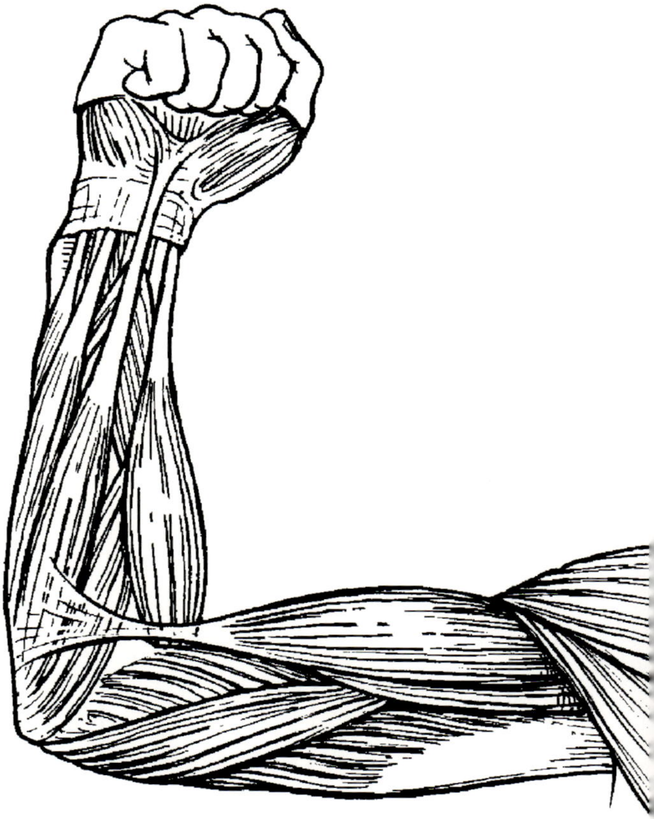

KEEPING THE MUSCULOSKELETAL SYSTEM AND SKIN HEALTHY

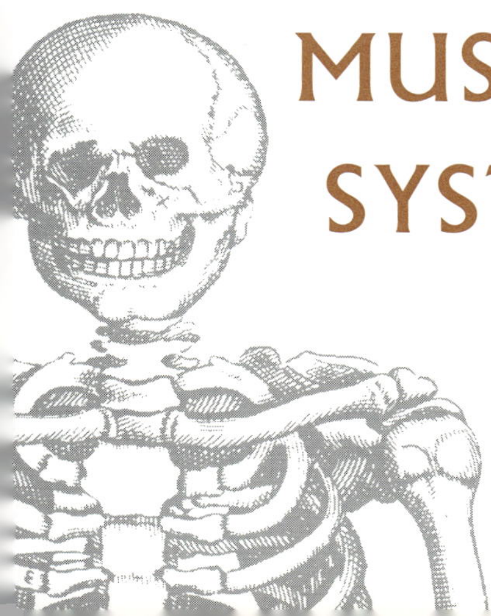

The human body is an amazing machine that works efficiently and effectively. People need to take care of their bodies to keep them working at their full potential. A healthy diet, plenty of exercise, and adequate rest are just some of the ways to make sure that the body is in the best shape possible.

EATING RIGHT FOR MUSCULOSKELETAL HEALTH

A good diet includes foods that are rich in vitamins, minerals, and other nutrients. This means plenty of fruits, vegetables, and other healthy foods. For children, whose bodies are still growing, protein is an important part of a healthy diet, as well. Proteins help build strong bones and healthy tissues.

There are plenty of foods that really help the musculoskeletal system. Cheese, milk, and eggs contain calcium and other important nutrients that strengthen bones and help them grow. Bread, rice, and pasta are good foods for muscles. They contain carbohydrates that supply the body with the energy that it needs to work effectively. Eating a proper diet can prevent problems in the future. For example, the risk of osteoporosis and other bone problems can be reduced by eating a diet that has plenty of protein and minerals.

THE ABCs OF EATING

Vitamins are an important part of a healthy diet. Here's how vitamins affect the musculoskeletal system.

- *Vitamin A* promotes the growth of both bone and cartilage. Foods rich in vitamin A include broccoli, carrots, dairy products, eggs, and spinach.
- *Vitamin C* helps form collagen. It is found in such citrus fruits as oranges, lemons, and limes, as well as tomatoes, potatoes, and strawberries.
- *Vitamin D* helps the body absorb calcium and phosphorous for strong, healthy bones. Foods rich in vitamin D include eggs, milk, and oily fishes. Vitamin D is also produced by skin cells when they are exposed to sunlight.
- *Vitamin E* contributes to the production of muscle tissue. This vitamin is found in nearly all foods.
- *Vitamin K* keeps bones healthy. Vitamin K is found in egg yolks, fish oils, liver, and green, leafy vegetables. Vitamin K is also made by some types of intestinal bacteria.

KEEPING FIT

Exercise is also important to staying healthy. It makes bones stronger. It is especially important for growing bodies. Exercise actually helps bones grow. Without exercise, the bones become weaker and less dense. They lose calcium and even begin to shrink.

Exercise also keeps muscles healthy and working properly. It improves muscle shape and tone. Without physical activity, muscles would *atrophy* (AA-truh-fee), or waste away.

For some people, lack of activity can cause real problems. Hospital patients, for example, and other people who are confined to bed rest for long periods of time, risk bone and muscle atrophy. Another group of people at risk of muscle atrophy and loss of bone density is astronauts. In space, astronauts lose calcium from the body, causing bones to shrink. Muscles are affected in space, too. There is no gravity exerting constant pressure on the muscles. Without use, the muscles atrophy. To stay fit while in space, astronauts must exercise a lot. Most astronauts exercise for at least two hours each day. The space shuttles are equipped with exercise equipment, such as treadmills, rowing machines, and stationary bicycles.

Staying hydrated is an important part of keeping healthy. The body naturally loses water through sweating. If sweat is not replaced with water, a person can get heatstroke and may even die. People are considered mildly dehydrated if they lose 3 to 5 percent of their body weight. Severe dehydration, resulting in a loss of 9 to 15 percent of body weight, is a medical emergency. Dehydration is more serious for children, who have smaller bodies than adults.

Professional athletes such as Tennessee Titans running back Eddie George know the benefits of staying hydrated while exercising, especially in warm weather.

PREVENTING INJURIES

Exercise and physical activity are important. Here are some tips to prevent sports-related or other injuries.

- Have an annual physical exam. Make sure that you talk to your doctor about any physical activities that you plan to do, especially if you have any health conditions that may be affected by exercise.
- Before exercising, it is important to warm up and stretch your muscles. If you don't stretch out, your body is more prone to cramps, sprains, and strains.
- Don't overdo it. Overexercising can cause problems, too. If muscles ache, give them a rest. Afterward, stretch again. After exercise, muscles tighten up and need to be loosened.
- *Never* play in pain or "through the pain." This can cause more damage. Also, avoid exercising or participating in sports when you're tired. Your reaction time is down when you're tired, and you are more prone to injury and mistakes.
- Stay hydrated! Drink lots of water while exercising, especially in hot weather. Your body loses water when you sweat.
- Clothes make the player. Shoes and shoe size can make a difference. The proper shoes can reduce the chance of such ankle injuries as sprains and strains. It is also important to replace sneakers and other athletic shoes when the treads wear out or the heels wear down. When playing sports, it is important to wear any safety equipment that the sport requires. Also, make sure that the equipment fits properly and is up to date.

Following are some other tips for keeping the musculoskeletal system safe and healthy.

- Watch the shape of things. Items that we use in the workplace and at home are now shaped to make our lives healthier and safer. Keyboards, office machines, chairs, and trackballs have all been redesigned to reduce musculoskeletal problems and the risk of RSIs.
- Stay drug-free. Avoid tobacco, alcohol, and other drugs that can adversely affect your health.
- Take a break. Rest and sleep are important to the body. When you sleep, your muscles get a chance to relax and rebuild. When you are tired, your reaction time is much slower.
- Keep it positive. A positive outlook on things can truly work wonders. Recent research has shown that upbeat people with arthritis experience less pain than people who are negative.

SKIN HEALTH

Protecting the skin from the sun is one of the most important ways to care for your skin. This includes wearing a sunscreen with an SPF of 15 or higher, putting on protective clothing, and staying out of the sun during peak hours. It is important to remember these tips when playing outside or participating in outdoor sports.

Eating well can keep skin healthy, too. Proteins, fresh fruits and vegetables, and plenty of liquids are all important. Foods rich in *riboflavin* and *niacin* can be especially beneficial to the skin. These nutrients can be found in cheese, eggs, fish, liver, meat, milk, and leafy vegetables.

Fast Fact

Did you know that you are taller when you sleep? That's because your muscles relax and stretch out.

Shape Up!

There are several different types of exercise. *Aerobic exercise* improves cardiac muscle and may also reduce joint inflammation associated with arthritis. This type of exercise includes anything that raises the heart rate for a sustained period, such as running, swimming, and cycling. *Stretching exercises* maintain regular joint movement and reduce the stiffness associated with arthritis. Activities that stress stretching include yoga and tai chi. Weightlifting and other *strengthening exercises* keep muscles strong and give them shape and definition.

A ballet instructor leads a group of young dancers through their exercise routine. Developing a pattern of regular exercise while young can lead to increased health later in life.

To ward off skin infections, acne outbreaks, and a serious case of the itches, keep the following tips in mind.

- Keep your hands clean. Even after washing, try not to touch your face in order to avoid spreading any bacteria that may be on the hands to the face.
- Wash your face twice a day with mild soap. Washing before bed is especially important. It removes the day's dirt and grime.
- Wash pillowcases and other bedding often to rid them of oil, dirt, and dead skin.
- For dry, itchy skin, use a moisturizing cream. A humidifier may also help to keep skin moist.
- Clean eyeglasses or sunglasses frequently to prevent oil from clogging pores around the nose.
- Don't scratch itchy skin. Scratching can break the skin and let in bacteria and other infection-causing materials.
- Remove makeup before you go to sleep.
- Drink a lot of water. This keeps the skin moist and flushes out wastes.
- Wear hair away from the face so that oil doesn't clog the pores.

Smoking and Wrinkles

One way to keep the skin looking as good as it can is to avoid smoking. Although researchers are not sure how, smoking does affect the skin. People who smoke have more wrinkles than people who don't smoke. Some health professionals believe that cigarette smoke may interfere with normal blood flow to the top layer of the skin. Others believe that smoking takes moisture out of the skin. Recent research has indicated that smoking may destroy collagen, which gives skin its elasticity.

GLOSSARY

appendicular skeleton—the bones that attach to the axial skeleton and permit movement and motion

arthritis—a musculoskeletal disorder characterized by painful stiffness and swelling in the joints

autoimmune disorder—a condition caused when the body's immune system attacks other parts of the body that it mistakenly identifies as foreign or harmful

axial skeleton—the bones that run lengthwise down the center of the body, giving it shape and structure

bacteria—microscopic, single-celled organisms that are found everywhere; many bacteria cause disease, but not all are harmful

benign—not cancerous

biopsy—a surgical procedure to remove tissue, cells, or fluids from the body for examination

birth defect—any physical problem with which an infant is born

bone marrow—a type of tissue found in the center of many long bones where infection-fighting cells are produced.

bone scan—a test of the skeleton in which a radioactive substance is injected into the bloodstream; afterward, the body is photographed by a special camera that can detect the radioactive substance

cancer—a disease in which a group of abnormal cells in the body divide without control

carcinogen—any substance that causes cancer

cardiac muscle—a special type of muscle found only in the heart

cartilage—strong, flexible connective tissue that covers bone ends

chemotherapy—the use of anticancer drugs to cure or stop the spread of cancers

compact bone—the hard, dense outer layer of bone

computerized tomography (CT)—a test that uses a thin beam of electromagnetic radiation to create a three-dimensional picture of what's inside the body

connective tissue—tissue that connects bones to other bones and muscles

contusion—a bruise caused by a blow to a muscle

cramps—painful, involuntary muscle contractions

dermatitis—any inflammation of the skin

dermis—the skin's middle layer

dislocation—the disconnection of the ends of two bones

electromyograph (EMG)—a test that uses electrical stimuli to detect muscle problems

environmental health—the body's reaction to conditions that humans have little or no control over, including sunlight, bacteria, chemicals, and pollution; it may also include medical problems that result from personal lifestyle choices, such as smoking and alcohol consumption

epidermis—the skin's paper-thin outer layer

fracture—any break in a bone

fungi—parasitic organisms, such as mushrooms and molds, that feed on both living and decaying matter

genetic disorder—an abnormality in a person's genes or chromosomes

hypodermis—the skin's deepest layer

joint—the place where two bones meet

ligament—strong tissue that connects one bone to another across a joint

magnetic resonance imaging (MRI)—a test that uses magnets and radio waves to create an image of the inside of the body

malignant—cancerous or harmful

marrow cavity—the space in the center of long bones that contains bone marrow

multiple chemical sensitivity (MCS)—a condition in which a person becomes allergic to many different chemicals after one initial episode of chemical exposure

muscular dystrophy—a genetic disorder that causes the body's skeletal muscles to weaken and deteriorate

musculoskeletal system—the body system that provides the framework for the body; it includes bones, muscles, joints, and connective tissues

osteoblasts—bone-building cells located in the end portions of long bones

osteoclasts—cells that break down bone

osteoporosis—a disorder that causes the loss of the minerals that make bone dense and strong

periosteum—a tough, thin protective membrane on the outside of bones

radiation—energy in the form of rays or particles

repetitive stress injury (RSI)—an injury that occurs when parts of the musculoskeletal system are stressed over and over; sometimes called overuse injury

skeletal muscles—muscles that can be moved under a person's control; also called voluntary muscles

skin—an organ that covers and protects the body's outer surface from contaminants; the skin also regulates body temperature and acts as a sensing organ

smooth muscles—muscles that power movements that are not under a person's control; also called involuntary muscles

spina bifida—a birth defect in which the backbone and the spinal canal do not close completely before birth

spongy bone—a type of bone, found inside compact bone, that is filled with many small spaces inside, just like a sponge; also called cancellous bone

synovial fluid—a watery liquid found in the joints that acts as a lubricant to allow ease of movement

synovial membrane—the capsule that surrounds movable joints and produces synovial fluid

tendon—a strong, elastic cord that connects muscle to bone or other muscle

toxin—poison

tumor—an abnormal mass of tissue that grows in the body; tumors may be benign or malignant

virus—a simple germ organism that is always harmful to humans; viruses are found everywhere

X-ray—a test that uses electromagnetic radiation to take a two-dimensional picture of what's inside the body

BIBLIOGRAPHY

BOOKS

Ballard, Carol. *The Skeleton and Muscular System.* Austin, TX: Raintree Steck-Vaughn, 1998.

Blakey, Paul. *The Muscle Book.* Honesdale, PA: Himalayan Institute Press, 2000.

Clayman, Charles, ed. *The American Medical Association Encyclopedia of Medicine.* New York: Random House, 1989.

The Human Body: An Illustrated Guide to Its Structure, Function, and Disorders. New York: DK Publishing, 1995.

Kittredge, Mary. *The Human Body: An Overview.* Philadelphia: Chelsea House, 2001.

Levy, Allan M. *Sports Injury Handbook: Professional Advice for Amateur Athletes.* New York: John Wiley and Sons, 1993.

Nagel, Rob, and Betz Des Chenes, eds. *Body by Design: From the Digestive System to the Skeleton.* Detroit: U*X*L, 2000.

Novick, Nelson Lee. *Skin Care for Teens.* Lincoln, NE: iUniverse.com, 2000.

Walker, Richard, ed. *Encyclopedia of the Human Body.* New York: DK Publishing, 2002.

WEB SITES

American Academy of Dermatology (AAD) www.aad.org

American Academy of Orthopaedic Surgeons (AAOS) www.aaos.org

CDC National Center for Environmental Health (NCEH) www.cdc.gov/nceh

The Children's Environmental Health Network www.cehn.org/cehn

The Institute for Preventative Sports Medicine (IPSM) www.ipsm.org

KidsHealth kidshealth.org

Muscular Dystrophy Association (MDA) USA www.mdausa.org

NIH National Institute of Arthritis and Musculoskeletal and Skin Diseases (NIAMS) www.niams.nih.gov/index.htm

NIH Osteoporosis and Related Bone Diseases—National Resource Center (ORBD—NRC) www.osteo.org

United States National Library of Medicine medlineplus.nlm.nih.gov

INDEX

Note: Page numbers in *italics* refer to maps, illustrations, or photographs.

AAP (American Academy of Pediatrics), 18
abscess, 84
Achilles tendon, 46, *50*
acne
- from air pollution, 104–105
- described, 97–98, *98*
- prevention of, 126

acne mechanica, 116
acrodynia, 70
aerobic exercise, 125
age, 15
Agent Orange, *68*, 68–69, 105
air pollution, *104*, 104–105
Akhenaton, 57
albinism, 93
Alliance to End Childhood Lead Poisoning, 65
alopecia areata, 96
ALS (amyotrophic lateral sclerosis), *38*, 38
amalgams, 70, *71*
American Academy of Pediatrics (AAP), 18
American Dental Association, 70
amyotrophic lateral sclerosis (ALS), 38, *38*
anemia, 24, 26
ankylosing spondylitis, 43
anorexia nervosa, 22
antagonistic muscle groups, 7
antibiotics, 82
Apert syndrome, 28–29
appendicular skeleton, 5, 127
arsenic, 62, *63*, 105
arthritis
- defined, 127
- Lyme disease and, 80, 82
- scleroderma as form of, 94
- symptoms of, 43
- treating, *44*, 44
- types of, 40–43, *41*

Arthritis Foundation, 44
aspirin, 44, 46
atopic dermatitis, *106*, 106–107
autoimmune disorders
- ankylosing spondylitis, 43
- defined, 127
- dermatomyositis, 56
- Marfan syndrome, 56–58, *57*
- myasthenia gravis, 36
- myositis, 36–37
- rheumatoid arthritis, 41
- skin disorders, 94–96, *95*

avascular necrosis, 47
axial skeleton, 5, 127

back injuries, 54
backpacks, 51
bacteria, bad, 80–86
- Lyme disease, 80–82, *81*
- osteomyelitis, 83–84, *84*
- Rocky Mountain spotted fever, 83
- sinusitis, *85*, 85
- skin conditions from, 117–118
- tetanus, 85–86

bacteria, defined, 127
ball-and-socket joints, *9*, 9
basal cell carcinoma, 111
basal cell layer, 89
Becker muscular dystrophy, 35
bending fracture, 15
benign, defined, 127
benign tumors, 25
Beriberi, 79
bibliography, 129
biological factors, vi
biopsy, 24, 127
birth defects, *27*, 27–29, 127
birthmarks, *92*, 92–93
blisters, 116, *117*
blood test, 37
BMD (bone mineral density) test, 23
bone biopsy, 24
bone cancers, 24–26
bone density, 22
bone disorders, 14–30
- birth defects and genetic disorders, *27*, 27–29
- bone cancers, 24–26
- conditions with unknown causes, 29–30
- fractures, 14–*21*, *16*, *19*, *21*
- osteoporosis, 22–24

bone grafting, 26
bone marrow, 3
bone mineral density (BMD) test, 23
bones
- composition, *3*
- exercise for, 122
- information from, *4*
- lead in, 64
- roles, 2
- types of, *6*

bone scan, 56, 127
bone spurs, 40
bone tumors, 25
books, 129
buckle fracture, 18
bunion, 58
bursae, 51
bursitis, 51–52

calcium, 23, 121, 122
callus, 17
cancellous bone, 3
cancer
- bone cancer, 24–26
- defined, 127
- skin cancer, 111–112, vi

candles, 64
carcinogen, 105, 127
cardiac muscle, 8, 127
cardiologist, 95
carpal tunnel syndrome, 52–53, *53*
cartilage
- arthritis and, 40
- defined, 127
- function of, *10*, 10, 12
- ossification and, 5
- types of, 12

cast, 17
Centers for Disease Control and Prevention (CDC), viii
chemical agents, viii
chemicals and disease, 60–73
- Agent Orange, *68*, 68–69
- arsenic, *63*
- exposure to, 60
- lead, *64*, 64–67, *66*
- mercury, 69–73, *71*, *72*
- sodium fluoride, 60–61, *61*
- top hazardous substances, 62–63

chemotherapy, 26, 127
chickenpox, 118
children
- bone health of, 5
- calcium deficiency of, 23
- growth plate injuries of, 19–20, *21*
- impetigo and, 109
- infectious diseases and, 118
- lead poisoning and, 64, 67
- mercury pollution and, 70
- repetitive stress injuries and, *53*, 53–54
- with rickets, 76
- at risk for environmental diseases, v
- sodium fluoride and, *61*, 61

chloracne, 105
chlorine, 103
chondrosarcoma, 25
chordoma, 25

clavicle (collarbone), 18
cleft lip, *27,* 27
cleft palate, 27, *27*
Clinton, Bill, 96
Clostridium tetani, 85
clothes, 123
clubfoot, 28
cold exposure, *115,* 115–116
collagen
in connective tissue, 10
function of, 2
scleroderma and, 95
smoking and, 126
Vitamin C for, 121
Colles's fracture, 18
comminuted fracture, 15–16
compact bone, 3, 127
compound fracture, 15
Comprehensive Environmental Response, Compensation, and Liability Act (CERCLA), 62
computerized tomography (CT), 55, 127
computers, *53,* 53–54
conditions with unknown causes
bone disorders, 29–30
joint and connective tissue disorders, 58
connective tissues. *See also* joint and connective tissue disorders
defined, 127
types of, *10,* 10–12, *11*
contusion
defined, 33, 127
example of, *32,* 32
cortical bone, 3
cortisone, 52
cramps, 32, 127
Crouzen syndrome, 27
CT (computerized tomography), 55, 127
cutis laxa, 93

da Gama, Vasco, 75
deciduous teeth, 6
deer ticks, 80–82
dehydration, 122
dentin, 6
deoxyribonucleic acid (DNA), 58
De Quervain's tenosynovitis, 50
dermatitis, *106,* 106–107, 127
dermatologist, 95, 98
dermatomyositis, 56
dermis, *88,* 89, 127
Derr, Pearl, 41, *41*
diagnosing
muscle disorders, 37, *37*
musculoskeletal problems, 55–56
diet
for arthritis, 44
for bone health, 5
leprosy and, 118
for musculoskeletal system health, 120–121, *121*
for myasthenia gravis, 36
for skin health, 124
diet-related conditions, 74–79
gout, *79,* 79
rickets, 76
scurvy, *74,* 74–75
spina bifida, 76–78, *77, 78*
disease. *See* chemicals and disease
dislocation, 18, 127
DM (myotonic dystrophy), 35
DMD (Duchenne muscular dystrophy), 35
DNA (deoxyribonucleic acid), 58
Doe Run Company, 67
"double-jointed" person, 10
drugs, 44, 56
Duchenne muscular dystrophy (DMD), 35

ear infection, 27
Ehlers-Danlos syndrome, 93
elastic cartilage, 12
elastin, 10
elderly people, 19
electricity, 17
electromyograph (EMG), 37, 127
ellipsoidal joint, *9,* 9
environmental agents, vi, 22
environmental diseases, vii
environmental health. *See also* musculoskeletal system, environment's effect on; skin, environment's effect on
defined, 127
factors that affect, vii, ix
Environmental Protection Agency (EPA), 67
epicondylitis, 51
epidermis, *88,* 89, 127
epidermolysis bullosa, 93
epiglottis, 12
estrogen, 22
Ewing's sarcoma, 25
exercise
for arthritis, 44, *44*
injury prevention, 123–124
for musculoskeletal system health, 122, *122,* 125, 125
for osteoporosis prevention, 23
for rheumatoid arthritis, 42
strains from, 45
facioscapulorhumeral muscular dystrophy (FSH), 35
FDA (Food and Drug Administration), 73
femur, 5
fibers, muscle, 7
fibrocartilage, 12
fibromyalgia, 42
fibrosarcoma, 25
fibula, *14,* 18
Fields, W.C., 96
fish, mercury poisoning from, 69, 72, 73
fixed joints, *9,* 9
flat bones, 5–6
fluoride, 60, 79
fluorosis, 61
folic acid, 76, 78
folliculitis, 116
food. *See* diet
Food and Drug Administration (FDA), 73
foot, tendonitis of, 50
forensic anthropologists, *4,* 4
formaldehyde, 105
fractures, 14–*21*
broken leg case, *16*
causes of, 14–15
common fractures, 18–19, *19*
defined, 127
fixing, 16–18
growth plate injuries, 19
Little League elbow, 20, *21*
osteoporosis and, 22
skateboard/scooter injuries, 18
of tibia and fibula, *14*
types of, 15–16
frostbite, *115,* 115–116
FSH (facioscapulorhumeral muscular dystrophy), 35
fungi, 83–84, 127

Gehrig, Lou, *38,* 38
genetic disorders
cleft palate, 27
defined, 127
Marfan syndrome, 56–58, *57*
muscular dystrophy, 33–35, *34*
skin disorders, *92,* 92–93
types of, 28–29
genetic tests, 37
George, Eddie, 122, *122*
German measles, 118
Gibson, J. Lockhart, 65
gliding joints, *9,* 9
glossary, 127–128
gout, 79, *79*
Grand Island YMCA's Arthritis Program, 44, *44*
Grand Rapids, MI, 60
greenstick fracture, 15
growth plate injuries, 19–20, *21*

hair follicles, *88,* 89, 96
Hansen's disease, 117–118
hazardous substances, 62–63, 63. *See also* chemicals and disease
head lice, 108, *108*
heart, 8
herbicide, 68
Herculaneum, MO, *66,* 67
hinge joint, *9,* 9
hip fracture, 19, *19*
houses, lead in, 65–66
human papillomavirus (HPV), 117
hyaline cartilage, 12
hydration, 122, 123
hydrocephalus, 78
hyoid bone, 6
hypodermis
  defined, 90, 127
  illustrated, *88*
hypothermia, 115

impetigo, 109
incontinentia pigmenti, 93
infections, 107, 109
infectious arthritis, 82
infestation, head lice, *108,* 108
inflammatory myopathy, 36–37
injuries. *See* repetitive stress injuries (RSIs); sports-related injuries
inorganic mercury, 69
involuntary muscles, 7
Iraq, mercury poisoning in, 72
irregular bones, 6

Jackson, Michael, 100
Japan, PCBs in, 104
joint and connective tissue disorders, 40–58
  arthritis, 40–44, *41, 44*
  autoimmune and genetic disorders, 56–58, *57*
  conditions with unknown causes, 58
  diagnosing, 55–56
  repetitive stress injuries, 49–54, *50, 53*
  sprains, strains, aches, pains, *45,* 45–48, *47*
joint aspiration, 56
joints
  defined, 127
  types of, 8–10, *9*
juvenile rheumatoid arthritis, 41

Karr, Megan, 71, *71*
keratinocytes, 89
Klippel-Feil syndrome, 28
Krause, Dr. Craig, 41, *41*
kyphosis, 29–30

landfills, lead in, 65
larynx, 12
lateral epicondylitis, 51
lead, 62, *64,* 64–67, *66*
lead choler, 65
leprosy, 117–118
lice, head, *108,* 108
lifestyle choices, vii
ligaments, *10,* 10, 127
Limb-girdle muscular dystrophy, 35
Lincoln, Abraham, 57, 58
Lind, James, 75
Little League elbow, 20, *21*
localized scleroderma, 94
long bones, 5, 19–20
lordosis, 30
"Lou Gehrig's disease," 38, *38*
lumbago, 46
lupus, 42, 43
Lyme disease, 80–82, *81*

Magellan, Ferdinand, 75
magnetic resonance imaging. *See* MRI (magnetic resonance imaging)
Maine, 70
malignant, defined, 127
malignant tumors, 25
malocclusion, 47
Marchment, Bryan, 32
Marfan syndrome, 56–58, *57*
marrow, 3, 127
marrow cavity, 3, 127
Mary Queen of Scots, 57
MCS (multiple chemical sensitivity), 73, 127
MD. *See* muscular dystrophy (MD)
measles, 118
medial epicondylitis, 51
melanin
  pigmentation disorders and, 93
  skin color and, 89
  sunburn and, 110
  UV rays and, 109
melanocytes, 99
melanoma
  causes of, 111–112
  chlorine and, 103
men, 43, 48
menopause, 22
mercury
  forms of, 69
  in the news, 70–71, *71*
  poisoning cases, 72, *72*
  pollution, 70
  safety tips, 73
  sources/effects of, 62
mercury salt, 69
metallic mercury, 69
metastatic bone cancer, 25
methylmercury
  poisoning, 71, 72
  release of, 69

Minamata disease, 71, *72,* 72
minerals, 2
mining, 70–71
Moeller, Dade W., 48
Mohan the elephant, *103,* 103
moles, 93
motor vehicle accidents, 48
MRI (magnetic resonance imaging), 45, 55, 127
MS (multiple sclerosis), 38
mucous membranes, 99
multiple chemical sensitivity (MCS), 73, 127
multiple sclerosis (MS), 38
muscle atrophy, 122
muscle disorders, 32–38
  autoimmune disorders, 36–37
  diagnosing, *37,* 37
  muscle injuries, *32,* 32–33
  muscle-related disorders, *38,* 38
  muscular dystrophy, 33–35, *34*
muscles
  exercise for, 122
  types of, *7,* 7–8
muscular dystrophy (MD)
  defined, 128
  described, 33–35
  example of, *34*
musculoskeletal system
  connective tissues, *10,* 10–12, *11*
  defined, 128
  diet for, 120–121, *121*
  environmental agents and, viii
  exercise for, 122, *122, 125,* 125
  injury prevention, 123–124
  joints, 8–10, *9*
  muscles, *7,* 7–8
  skeleton, 2–6, *3, 4, 6*
musculoskeletal system, environment's effect on, 60–86
  bad bacteria, 80–86, *81, 84, 85*
  chemicals/disease, 60–73, *61, 63, 64, 66, 68, 71, 72*
  diet-related conditions, *74,* 74–79, 77, 78, 79
myasthenia gravis, 36
myeloma, 26
myelomeningocele spina bifida, 78
myocardium, 8
myositis, 36–37
myotonic dystrophy (DM), 35

National Institute of Arthritis and musculoskeletal and Skin Diseases (NIAMS), 22, 42
National Institute of Environmental Health Sciences (NIEHS), ix
National Institutes of Health (NIH), 23
neck conditions, 48
neural tube defect (NTD), 77

niacin, 124
"Nintendo thumb," 53
non-Hodgkin's lymphoma, 26, 68
NTD (neural tube defect), 77
nutritional deficiency, vi. *See also* diet

obesity, 99
occupational disorders, 48. *See also* repetitive stress injuries (RSIs)
ocean pollution, 102
oil glands, *88,* 89
Oliver, Dr. James, 81, *81*
organic mercury, 69
Osgood-Schlatter disease (OSD), 54
ossification, 5
osteitis deformans, 30
osteoarthritis, 40
osteoblasts
- defined, 5, 128
- for healing fracture, 16

osteoclasts, 5, 128
osteogenesis imperfecta, 28
osteomyelitis (bone infection)
- with compound fracture, 15
- described, 83–84, *84*

osteoporosis
- bone fractures and, 15
- calcium deficiency and, 79
- defined, 22, 128
- lead exposure and, 64
- risk factors, 24
- sodium fluoride link to, 61
- treating, 23

osteosarcoma, 25
overuse injuries. *See* repetitive stress injuries (RSIs)
ozone layer, 110

Page, Bobby, *57,* 57
Paget's disease, 30
paralysis, 19
Pauling, Linus, 75
PCBs (polychlorinated biphenyls), 63, 104–105
periosteum, 3, 128
permanent teeth, 6
Perthes disease, 30
Phimthep, Thawiphop, 27, *27*
phototherapy, 94
physical agents, viii
physical anthropologists, 4
Pierre Robin sequence, 28
pigmentation disorders, 93
pigmented birthmarks, 93
pivot joints, 9, *9*
plasma, 26
polio, 38
pollution. *See also* chemicals and disease
- affect on skin, 102–105, *103, 104*
- environmental health and, vii

polychlorinated biphenyls (PCBs), 63, 104–105
polyvinyl chloride (PVC), 62
porphyria cutanea tarda, 105
posture, 54
pregnancy
- bone lead and, 64
- mercury poisoning and, 73
- spina bifida and, 78

prenatal testing, 93
primary bone cancer, 25
psoriasis, 94
pus, 84
PVC (polyvinyl chloride), 62

radiation
- defined, 128
- treatment for bone cancer, 26
- from x-ray/CT machine, 55

red bone marrow, 3
Rembrandt, Harmenszoon van Rijn, 96
remission, 42
repetitive stress injuries (RSIs), 49–54
- back injuries, 54
- bursitis, 51–52
- carpal tunnel syndrome, 52–53
- children and, *53,* 53–54
- defined, 128
- Osgood-Schlatter disease, 54
- symptoms/causes of, 49
- tenosynovitis/tendonitis, *50,* 50–51

rheumatoid arthritis, 41–42
riboflavin, 124
RICE (rest, ice, compression, elevation), 46
rickets, 76
Rocky Mountain spotted fever (RMSF), 83
rosacea, 96–97, *97*
rotator cuff tendonitis, 50–51
RSIs. *See* repetitive stress injuries (RSIs)
rubella, 118
rubeola, 118
Rucinsky, Martin, 32, *32*

saddle joints, 9, *9*
sailors, 74, 75
scabies, 107
sciatica, 48
scleroderma, 94–95, *95*
scoliosis, 29
scooter injuries, 18
scurvy, *74,* 74–75
sea lice, 102
seborrhea, 99
secondary bone cancer, 25
short bones, 5

simple fracture, 15
sinusitis, *85,* 85
skateboard injuries, 18
skeletal muscles
- defined, 128
- function of, 7
- myasthenia gravis and, 36

skeleton
- bone composition, 2–3, *3*
- bone roles, 2
- information from, 4
- types of bones, 5–6, *6*

skin
- defined, 128
- effect of environmental agents, viii
- functions of, 88
- health, 124–126
- layers of, *88,* 89–90

skin, environment's effect on, 102–118
- dermatitis, *106,* 106–107
- infections and infestations, 107–109, *108*
- pollution in air, *104,* 104–105
- pollution in water, 102–103, *103*
- sports and, 116–117, *117*
- viruses and bad bacteria, 117–118
- weather, 109–116, *113, 115*

skin cancer
- cause of, viii
- types of, 111–112

skin disorders, 92–100
- autoimmune disorders, 94–96, *95*
- genetic disorders, *92,* 92–93
- miscellaneous conditions, 96–100, *97, 98, 99*

skin graft, 100, 112
smoking, 115, 126
smooth muscles, 7, 128
sodium fluoride, 60–61, *61*
spasms, 32
spina bifida
- Agent Orange and, 68
- case of, *77*
- cause/types of, 76–78
- defined, 128
- rate of babies born with, *78*

spinal conditions, joint, 48
spinal curvature, 29
spinal fracture, 19
spongy bone, 3, 128
sports, affect on skin, 116–117, *117*
sports-related injuries
- common, 55
- growth plate injuries, 19
- Little League elbow, 20, *21*
- prevention of, 123–124
- strains, 45–46

sprains, 45, 46–47
squamous cell carcinoma, 111
squamous cell layer, 89

stapes, 5
staphylococcus, 84
stenosing tenosynovitis, 51
Stepanek, Mattie, *34,* 34
strains
  causes/symptoms of, 45–46
  defined, 33
  treating, 46–47
strengthening exercises, 125
Streptococcus pneumoniae, 85
stress fracture, 14, 15
stretching exercises, 125
sunburn, 110–111, 113, *113*
sunglasses, 113, 114
sunlight
  damage from, 109
  protection from, *113,* 113–114, 124
  skin cancer from, 111–112
  sunburn, 110–111
sunscreen, 110, 113, 114, 124
sweat glands, *88,* 90
synovial fluid
  defined, 8, 128
  joint cracking and, 11
synovial membrane, 8, 128
systemic lupus erythematosus, 42, 43
systemic scleroderma, 94–95
Szent-Györgyi, Albert, 75

tans, 111
tears, muscle, 33
teeth, 6, *6*
temporomandibular joint (TMJ) disorder, 47, *47*
tendonitis, *50,* 50–51
tendons
  defined, *10,* 10, 128
  tenosynovitis/tendonitis, *50,* 50–51
tennis elbow, 51
tenosynovitis, *50,* 50–51
tetanus, 85–86
thiamine, 79
tibia, *14,* 18
ticks, 80–83, *81*
tinea infections, 116–117
TMJ (temporomandibular joint) disorder, *47,* 47
tooth decay, 79
toxin, 128
Treacher-Collins syndrome, 28
trigger finger, 51
"Trucker's back," 54
tumor, 24, 25, 26, 55, 112, 128

ultraviolet (UV) rays
  ozone layer and, 109
  protection from, 114
  skin cancer and, 111–112
  for vitamin D, 76
ultraviolet-A (UVA) rays, 109
ultraviolet-B (UVB) rays, 109
uric acid crystals, 79
UV index, 114

vaccine, 86, 118
varicella, 118
vascular birthmarks, *92,* 93
video games, *53,* 53–54
Vietnam War, 68
vinyl chloride, 62–63
viruses, 117–118, 128
vitamin A, 121
vitamin C, 74, 75, 121
vitamin D
  deficiency, rickets from, 76
  effect on body, 121
  for osteoporosis prevention, 23
  from sunlight, 109
vitamin E, 121
vitamin K, 121
vitamins, 120–121
vitiligo, *99,* 99–100
voluntary muscles. *See* skeletal muscles

warts, 117
water pollution, 102–103, *103*
weather, 109–116
  skin cancer, 111–112
  sunburn, 110–111
  sunburn protection, *113,* 113
  UV rays, 109–110
  UV rays, protection from, 114
  wind and cold, *115,* 115–116
Weathers, Dr. Seaborn Beck, 115, *115*
Web sites, 129
Wells, David, *79,* 79
whiplash, 48
wind exposure, 115–116
wisdom teeth, 6
women, 22, 41, 42
workplace, v
wrinkles, 126
wrist drop, 65
wrist ganglion, 58

xeroderma pigmentosum, 93
X-ray, 55, 128

Yamuna River, pollution of, *103,* 103
yellow bone marrow, 3
Yoder, Justin, *77,* 77
Yoder, Myron, 77, *77*
Yusho (oil disease), 104